GCSE Additional Applied Science

Contents

Welcome to AQA GCSE Science!

Key points

At the start of each topic are the important points that you must remember.

This book has been written for you by the people who will be marking your exams, very experienced teachers and subject experts. It covers everything you need to revise for your exams and is packed full of features to help you achieve the very best that you can.

Key words are highlighted in the text and are shown **like this**. A glossary of these terms can be found within Kerboodle, www.kerboodle.com

▸ *These questions check that you understand what you're learning as you go along. The answers are all at the back of the book.*

Many diagrams are as important for you to learn as the text, so make sure you revise them carefully.

Anything in the Higher boxes must be learned by those sitting the Higher Tier exam. If you're sitting the Foundation Tier, these boxes can be missed out.

The same is true for any other places that are marked [**H**].

Higher

AQA Examiner's tip

AQA Examiner's tips are hints from the examiners who will mark your exams, giving you important advice on things to remember and what to watch out for.

Bump up your grade

How you can improve your grade – this feature shows you where additional marks can be gained.

Maths skills

This feature highlights the maths skills that you will need for your Science exams with short, visual explanations.

At the end of chapters 2–5 you will find:

End of chapter questions

These questions will test you on what you have learned throughout the whole chapter, helping you to work out what you have understood and where you need to go back and revise.

AQA Examination-style questions

These questions are examples of the types of questions you will answer in your actual GCSE, so you can get lots of practice during your course.

You can find answers to the End of chapter and AQA Examination-style questions at the back of the book.

Student Book
pages 2–7

1.1–1.3 Health, hazards, risks and fire

- The **Health and Safety Executive** (HSE) is responsible for the regulation of risks to health and safety in the workplace.

▶ **1** *Why do employers need the cooperation of employees to ensure their safety?*

- Safety signs and **hazard** symbols are easy to recognise once you know what they look like.

| electrical | radioactivity | no smoking | wear eye protection | first aid | highly flammable |

biohazard   toxic   harmful/irritant   corrosive   explosive   oxidising   dangerous for the environment

Safety signs and hazard symbols

▶ **2** *How can you recognise signs that: (a) warn of danger, (b) tell us what we 'must not do', (c) tell us what we 'must do', (d) give safety information, (e) show hazard symbols?*

- In all activities check your work space for potential dangers. Follow **risk assessments** to control hazards and limit the **risk** to yourself and others. Stay alert to problems and report incidents. Always keep risks As Low As Reasonably Practicable (ALARP).

▶ **3** *What is the difference between a hazard and a risk?*

- Different extinguishers should be used to put out different types of fire.

Water	Powder	Foam	Carbon dioxide (CO$_2$)
For wood, paper, textiles and solid material fires	For liquid and electrical fires	For use on liquid fires	For liquid and electrical fires
DO NOT USE on liquid, electrical or metal fires	**DO NOT USE** on metal fires	**DO NOT USE** on electrical or metal fires	**DO NOT USE** on metal fires

Types of fire extinguisher

▶ **4** *Which extinguisher is best for a fire in (a) a wood store, (b) a computer repair shop?*

Key words: Health and Safety Executive, hazard, risk assessment, risk

Key points

- The Health and Safety at Work Act puts the responsibility for safety on both employers and employees.
- Biological, chemical and physical hazards exist in science labs.
- A hazard is a danger; a risk is the chance of being harmed.
- A risk assessment helps you to control hazards and limit risks.
- There are four main types of fire extinguisher: water, powder, foam and carbon dioxide. They are used on different types of fire.

AQA Examiner's tip

Knowing the safety signs and hazard labels is an easy way to pick up marks in your examination.

Student Book
pages 8–9

1.4 Following standard procedures

- Scientists follow **standard procedures** to make sure the materials and products they investigate, or make, are safe and reliable in use. In experimental work, standard procedures help scientists to avoid **errors**. They also enable other scientists to **repeat** the same experiment and **reproduce** the results.

Key points

- A standard procedure is an agreed way of doing something.
- Scientists follow standard procedures to obtain repeatable and reproducible results.
- The scientific method is a set way of carrying out scientific research.

|||▶ **1** *Why do scientists follow a standard procedure to test the quality of drinking water?*

- The **scientific method** is an approach (or series of steps) scientists can use in their research.

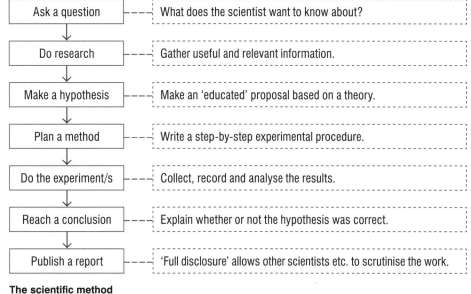

Ask a question	What does the scientist want to know about?
Do research	Gather useful and relevant information.
Make a hypothesis	Make an 'educated' proposal based on a theory.
Plan a method	Write a step-by-step experimental procedure.
Do the experiment/s	Collect, record and analyse the results.
Reach a conclusion	Explain whether or not the hypothesis was correct.
Publish a report	'Full disclosure' allows other scientists etc. to scrutinise the work.

The scientific method

Bump up your grade

To improve your mark, you should not only know the order of the steps in a scientific method, but what those steps mean.

|||▶ **2** *What is a hypothesis?*

Key words: standard procedure, error, repeat, reproduce, scientific method

Student Book
pages 10–11

1.5 Carrying out a standard procedure

- When monitoring and controlling processes, and when making and analysing substances, scientists need results they can depend on. Standard procedures ensure everyone carries out a particular task in exactly the same way. This is to ensure that results can be repeated and are comparable. They can be reproduced by other scientists in different organisations.

Key points

When following a standard procedure you should:

- first check you understand it
- set out your work area safely and tidily
- follow the instructions step-by-step
- keep alert to sources of error and repeat observations.

|||▶ **1** *How do standard procedures help to make results repeatable and reproducible?*

Student Book
pages 14–15

2.1 Your heart

- Your right **ventricle** pumps **deoxygenated blood** to your lungs.
- Oxygenated blood returns to your left **atrium**.
- Your left ventricle pumps **oxygenated blood** to the rest of your body.
- Deoxygenated blood returns to your right atrium.

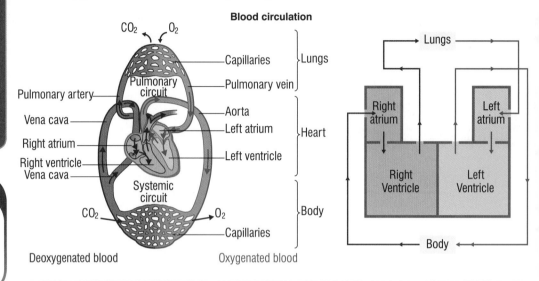

Blood circulation

IIIII➡ 1 *What are the lower chambers of the heart called?*

- Heart rate is a **baseline measurement**. We measure heart rate in beats per minute. You can measure heart rate by counting the number of pressure pulses in 30 seconds, then multiply by 2.

Key words: ventricle, deoxygenated blood, atrium, oxygenated blood, baseline measurement

Key points

- Your cardiovascular system is your heart and its blood vessels.
- Blood leaves your heart's ventricles through arteries. Blood returns to your heart's atria through veins.
- Heart rate is a baseline measurement that changes during exercise.

AQA Examiner's tip

Note that the only vein carrying oxygenated blood is the pulmonary vein, which goes from your lungs to your heart.

Student Book
pages 16–17

2.2 Your lungs

- When you breathe in your lungs increase in size because:
  - your **diaphragm** contracts and moves down.
  - your **intercostal muscles** contract and lift your ribs up and out.
- When the muscles in your **thorax** relax you are breathing out.

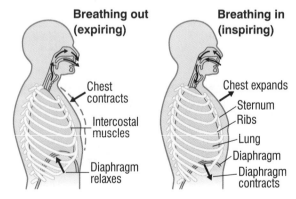

Breathing in increases the size of your thorax

IIIII➡ 1 *What does your diaphragm do to inflate your lungs?*

- You measure breathing rate by counting the breaths in one minute.
- Use a spirometer to measure (in cm^3) your:
  tidal volume (TV) – the volume of air you breathe out normally
  vital capacity (VC) – the maximum volume of air you can force out.

IIIII➡ 2 *What is the difference between TV and VC?*

Key points

- Your diaphragm and the intercostal muscles between your ribs help you to breathe.
- Breathing rate, tidal volume and vital capacity are often taken before and after exercise.

AQA Examiner's tip

The movement of your diaphragm and ribcage are controlled by muscles.

Key words: diaphragm, intercostal muscle, thorax, tidal volume, vital capacity

Student Book
pages 18–19

Key points

- Your heart and lungs provide glucose and oxygen (dissolved in blood) for your muscles.
- Aerobic respiration needs oxygen. Anaerobic respiration does not need oxygen.
- Anaerobic respiration takes over when an oxygen debt occurs.

AQA *Examiner's tip*

Respiration is not breathing in and out. Respiration is the process of releasing energy from glucose.

2.3 Changes during exercise

- **Aerobic respiration** occurs during light exercise such as jogging.

$$\text{glucose} + \text{oxygen} \longrightarrow \text{carbon dioxide} + \text{water} \ (+ \text{energy})$$
$$C_6H_{12}O_6 + 6O_2 \longrightarrow 6CO_2 + 6H_2O \ (+ \text{energy})$$

➡ **1** *What happens to your breathing rate and heart rate during exercise?*

➡ **2** *Write the word equation for aerobic respiration.*

- **Anaerobic respiration** occurs if insufficient oxygen is reaching your muscles.

$$\text{glucose} \longrightarrow \text{lactic acid} \ (+ \text{energy})$$

Anaerobic respiration releases much less energy than aerobic respiration. When you have an **oxygen debt** the build-up of **lactic acid** makes your muscles ache. This lactic acid must be turned into glucose.

➡ **3** *What is oxygen debt?*

Key words: aerobic respiration, anaerobic respiration, oxygen debt, lactic acid

Higher

Student Book
pages 20–21

Key points

- Your body cannot store glucose in your blood. Instead you store it as a starch, called glycogen, in your liver (20%) and your muscles (80%).
- The hormones insulin and glucagon [H] control your blood glucose levels.

AQA *Examiner's tip*

When you revise make simplifying notes containing the key words; glycogen, insulin, glucose and glucagon.
Think of ways to avoid muddling the key words:

- There is no 'e' in insulin and glucagon. These are the hormones.
- There is an 'e' in glycogen and glucose. These are the energy store and fuel.

2.4 Recovery after exercise

- Digesting carbohydrate foods puts **glucose** into your blood. When there is too much glucose in your blood, your pancreas secretes **insulin**. Insulin makes your liver convert soluble glucose into **glycogen**, which is then stored.

➡ **1** *Which hormone is in control when there is too much glucose in your blood?*

➡ **2** *What is glucose converted to for storage in the liver?*

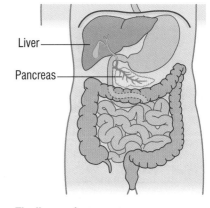

Liver

Pancreas

The liver and pancreas

Exercise removes glucose from your blood. When there is too little glucose in your blood, your pancreas secretes the hormone **glucagon**. Glucagon makes your liver convert glycogen into glucose, which is then released into your bloodstream.

➡ **3** *What happens to blood glucose levels during exercise?*

Key words: insulin, glycogen, glucagon

Higher

**Student Book
pages 22–23**

2.5 Controlling temperature and fluid levels

Key points

- During exercise your core body temperature increases. You sweat and blood capillaries below your skin carry more blood.
- You remove water in your body in sweat and urine.

Bump up your grade

During exercise your skin does not go red because you are sweating. Your skin is redder because blood vessels allow blood to flow nearer the surface in the capillaries. Heat leaves your body due to evaporation (sweating) and radiation (from blood in the capillaries).

- Your body works best when the conditions inside it (the internal environment) remain the same.

> **1** *Why does your body need to maintain a core temperature of 37°C and steady blood water levels?*

- Heat loss by **evaporation** and by **radiation** cause cooling:
  - When you get too hot your **sweat glands** release sweat. The sweat takes energy from your skin as the water in sweat evaporates.
  - When you get too hot the blood vessels supplying your **capillaries** just below your skin expand (dilate). More blood flows near the surface. Hotter skin radiates heat away more quickly.
- Your sweat glands and **kidneys** control your water output. Kidneys remove excess water into your urine. You store urine in your **bladder** until you empty it in the toilet.

> **2** *What glands in your skin help control both temperature and water levels?*

> **3** *What happens in your skin during exercise?*

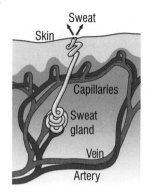

Your skin helps to control your temperature

Key words: evaporation, radiation, sweat gland, capillary, kidney, bladder

**Student Book
pages 24–25**

2.6 Physiotherapy

Key points

- Physiotherapists massage muscles and manipulate joints when treating injured athletes.
- Your skeleton supports your body and protects your vital organs.
- Your muscles allow your body to move.

AQA Examiner's tip

Friction either causes objects to grip or slip. Cartilage and synovial fluid are slippery, so reduce friction.

- Common injuries include damaged **ligaments**, pulled or torn **muscles**, torn **cartilage**, ruptured **tendons**, dislocated **joints** and fractured **bones**.
- **Tendons** attach muscle to bone – this enables bones to move.
- **Ligaments** link bones – this stabilises joints.

> **1** *What is the difference between a tendon and a ligament?*

Cartilage and **synovial fluid** prevent friction between bones at synovial joints such as the knee.

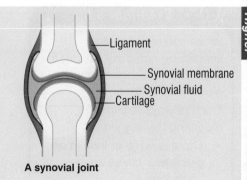

A synovial joint

> **2** *What is the lubricating fluid produced by the synovial membrane?*

Key words: ligament, muscle, cartilage, tendon, joint, synovial fluid

**Student Book
pages 26–27**

2.7 Biomechanics – the science of human movement

Key points

- Skeletal muscles work as antagonistic pairs. They pull on bones to make them pivot.
- A moment is the turning effect of a force.

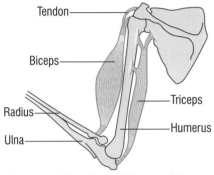

Tendon

Biceps

Triceps

Radius

Humerus

Ulna

An antagonistic pair – the biceps and triceps

Maths skills

Calculating a turning effect or moment:

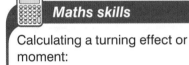

Force = 40 N   90°   Pivot
Distance = 0.3 m

Moment = force × distance
= 40 N × 0.3 m
= 12 N m

- A muscle can only pull on a bone, it cannot push. For this reason muscles always work in **antagonistic pairs**, as one muscle contracts the other relaxes. Your **biceps** and **triceps** make your arm bend at your elbow.

▶ **1** *What happens to your biceps and triceps when you extend your forearm?*

- You calculate a **moment** using the equation:
  Moment = force × perpendicular distance to the pivot
  Moments have units of newton metre or N m.

▶ **2** *The moment exerted to lift a shot-putt is 2.4 N m. If the force exerted by the biceps is 80 N, what is the distance B between the elbow and the biceps?*

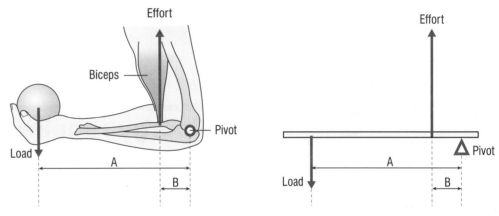

Forearm as a simple lever. A and B are the distances from the pivot to where the forces act.

Key words: antagonistic pair, moment

**Student Book
pages 28–29**

2.8 Prosthetics

Key point

- Apart from being comfortable, artificial joints must be lightweight and strong.

Bump up your grade

Be prepared to discuss the appropriateness of certain materials for artificial joints and issues arising from their use.

- Titanium alloy is the best lightweight, yet strong, metal for artificial joints.
- Plastic polymers can be moulded to joints. Kevlar fibres add reinforcement. Polyethene foam makes a comfortable soft-socket pad for amputees.

▶ **1** *Why does 'Flex-Foot' need to be light, strong and elastic?*

Oscar Pistorius, the 'Blade Runner' on his Flex-Feet

Student Book
pages 30–31

2.9 Nutrition for exercise and fitness

- A balanced diet contains the **nutrients**: carbohydrates, fats, proteins, vitamins, minerals and water.
- **Carbohydrates**, from sugar and starch, break down into glucose which your muscles use for energy. You store glucose as glycogen in your liver and muscles (See 2.4 Recovery after exercise).

▶ **1** *Which carbohydrate does your body use to obtain energy by respiration?*

- **Proteins,** from meat, fish, dairy products, cereals and beans, aid the growth and repair of body tissues (such as muscle).
- You can use a **diet diary** to track how many kilojoules (kJ) of energy you eat from each source of food. (1 kcal = 4.2 kJ)

▶ **2** *Why shouldn't you leave more than 24 hours between eating and recording information in a sports' diet diary?*

Key words: nutrient, carbohydrate, protein, diet diary

Key points

- Only with a correct intake of nutrients can an athlete reach their full potential.
- The simplest way to monitor your diet is to keep a diet diary.

⚖	NUTRITION		
TYPICAL VALUES	PER SHEET (APPROX 16 g)	PER 100 g (UNCOOKED)	
Energy Value (calories)	230 kJ (55 kcal)	1470 kJ (345 kcal)	
Protein	2 g	12 g	MEDIUM
Carbohydrate (of which sugars	12 g 0.3 g	72 g 2 g)	HIGH LOW
Fat (of which saturates	0.2 g Trace g	1 g 0.3 g)	LOW LOW
Fibre	0.5 g	3 g	MEDIUM
Sodium	Trace g	Trace g	LOW

Nutrition label on a packet of lasagne

Student Book
pages 32–33

2.10 Energy requirements

- Your daily energy requirement depends on your body mass and level of exercise.
- **Basic Energy Requirement (BER)** is 5.4 kJ/hour for every kilogram of body mass.

▶ **1** *What is the basic energy requirement of a 50 kg athlete each day?*

- **Body Mass Index (BMI)** has units of kg/m².

$$\text{BMI} = \frac{\text{mass in kg}}{(\text{height in m})^2} = \frac{m}{h^2} = \frac{m}{h \times h}$$

▶ **2 a** *A student has a mass of 80 kg and a height of 1.8 m. Calculate his BMI.*
  b *A BMI of 25 or more normally indicates someone is overweight. What would you say to a student with a BMI of 30?*

▶ **3** *A fit, short, muscular weightlifter has a BMI of over 28. Why doesn't this BMI indicate a problem?*

Key words: Basic Energy Requirement (BER) , Body Mass Index (BMI)

Key points

- All athletes need a carbohydrate-rich diet that provides enough energy.
- Your body needs more than a basic amount of energy if you exercise.
- Body mass index (BMI) is a good indicator of total body fat and reveals your ideal weight, based on your height.

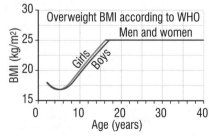

Graph showing BMI information from the World Health Organisation

AQA *Examiner's tip*

Do not rush a calculation about basic energy requirement (BER). Substitute, calculate and remember to put the correct units of kJ at the end.
BER = kg × 5.4 kJ × hours

Bump up your grade

When you calculate BMI follow the steps: *Formula, substitute, calculate, round off answer, add correct units.*

**Student Book
pages 34–35**

2.11 Sports drinks and sports diets

- Water in an **isotonic** sports drink replaces fluid lost as sweat during exercise, while glucose provides an energy boost to the athlete. **Electrolytes**, such as sodium chloride, increase the absorption of the drink from the intestine into the blood and reduce urine output.
- Sports drinks are **hypertonic** – high in glucose, or **hypotonic** – low in glucose.

> **1** *Why do isotonic sports drinks contain: (a) water (b) glucose (c) sodium chloride?*

> **2** *Which type of sports drink is better to boost your energy levels – hypertonic or hypotonic?*

- Athletes need to have a **carbohydrate-rich** diet to increase their glycogen stores.
- Some power athletes need a **protein-rich** diet in order to build muscle.

> **3** *Why do sprinters need a high-protein diet?*

Key points

- Isotonic sports drinks contain water, glucose and electrolytes.
- All athletes need a carbohydrate-rich diet that provides enough energy.
- Power athletes, such as sprinters, train to gain muscle mass. They have a high-protein diet.

Isotonic sports drinks

AQA Examiner's tip

The high-protein diet of a power athlete is also carbohydrate-rich.

Key words: isotonic, electrolyte, hypertonic, hypotonic

**Student Book
pages 36–37**

2.12 Standard procedures for maintaining health and fitness

- Doctors, nurses and **pharmacists** help prevent, diagnose and treat medical conditions.
- **Nutritionists** and **dieticians** encourage us to stay fit and healthy.
- **Physiologists** and training coaches deal with parts of the body involved in exercise.
- **Physiotherapists** specialise in muscle problems caused by accidents, illness and ageing.

> **1** *State two healthcare occupations and the roles of these scientists.*

- You measure blood-glucose levels using a **glucose-testing strip**, which changes colour based on the concentration of glucose in urine.
- You squeeze a handgrip **dynamometer** to measure muscle strength.

> **2** *How could swinging a dynamometer to squeeze it tighter give an unreliable handgrip strength test?*

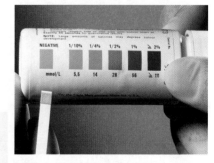
Diastix glucose-testing strip

Key points

- A range of occupations exist in healthcare.
- Scientists must make reproducible and accurate measurements to monitor the physiological changes of people during exercise.
- A standard procedure is like a recipe – an agreed way of doing something.

AQA Examiner's tip

Check you also understand the other standard procedures in this topic: *pulse* (2.1), *breathing rate, VC, TV* (2.2) and *body temperature* (2.5), and how to calculate a *moment* (2.7), *BER* and *BMI* (2.10).

Key words: dietician, physiotherapist, glucose-testing strip

1 Why do all athletes need a high carbohydrate diet?

2 What happens to your biceps and triceps when you bend your elbow?

3 Which vein carries oxygenated blood from your lungs to your heart?

4 What is the name of the hormone that helps your liver convert glycogen to glucose when there is too little glucose in your blood? [H]

5 What are the three main ingredients of isoelectronic sports drinks?

6 What is the basic energy requirement (BER) of an 80 kg athlete for one day, if BER = 5.4 kJ per hour for every kg of body mass?

7 What muscles allow your ribcage to expand and contract?

8 What is the function of

    a tendons
    b ligaments?

9 How does your skin help to control your body temperature?

10 What is oxygen debt? [H]

Chapter checklist ✓✓✓

Tick when you have:

	✓	✓	✓
reviewed it after your lesson	✓		
revised once – some questions right	✓	✓	
revised twice – all questions right	✓	✓	✓

Move on to another topic when you have all three ticks

	✓	✓	✓
Your heart	✓	✓	✓
Your lungs	☐	☐	☐
Changes during exercise	☐	☐	☐
Recovery after exercise	☐	☐	☐
Controlling temperature and fluid levels	☐	☐	☐

	✓	✓	✓
Physiotherapy	☐	☐	☐
Biomechanics – the science of human movement	☐	☐	☐
Prosthetics	☐	☐	☐
Nutrition for exercise and fitness	☐	☐	☐
Energy requirements	☐	☐	☐
Sports drinks and sports diets	☐	☐	☐
Standard procedures for maintaining health and fitness	☐	☐	☐

1 To ensure optimum performance, athletes follow carefully controlled diets and strict exercise routines.

   a Match the following dietary components to their role in the body. *(3 marks)*

Dietary component	Role in the body
Carbohydrates	Used for growth and repair
Proteins	Maintain health
Vitamins and minerals	Act as a store of energy, and are used for insulation
Fats	Main source of energy

   b Athletes generate energy through the process of respiration. Write the word equation for respiration. *(2 marks)*

   c Describe how the blood glucose levels of an athlete would vary after eating, and during exercise. *(2 marks)*

   d *In this question you will be assessed on using good English, organising information clearly and using specialist terms where appropriate.*

     Explain, by referring to specific hormones, how the body ensures that an athlete's blood glucose levels remain constant. *(6 marks)*

2 Physiologists and training coaches deal with parts of the body involved in exercise. To help them study how the body changes during exercise, a range of baseline measurements are taken.

   a Taking a person's resting pulse rate is an example of a baseline measurement. What are the units used to measure pulse rate? *(1 mark)*

   b Other than pulse rate suggest **two** changes which take place when a person exercises, which can be detected by changes in baseline measurements *(2 marks)*

   c State and explain **two** changes which allow the body to control its temperature during exercise. *(4 marks)*

   d During exercise blood has to flow through the body more quickly, to ensure muscle cells are supplied with oxygen and glucose. Describe how blood flows through the heart to deliver oxygenated blood to the cells. *(4 marks)*

3 Physiotherapists help people to regain mobility after injury or surgery.

   a Suggest **two** common injuries that a physiotherapist may treat. *(2 marks)*

   b Copy and complete the following sentences, choosing the correct term from below:

     pulling   contracting   relaxing   pushing

     Muscles cause movement in the skeleton by _____ on bones. They do this by _____ *(2 marks)*

   c A goalkeeper's arm can be modelled as a lever, as follows:

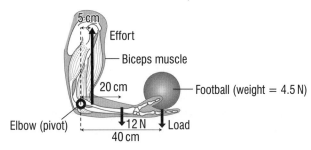

    i Calculate the moment of the ball about the elbow. *(3 marks)*

    ii What is the moment about the elbow, caused by the weight of the arm? *(1 mark)*

    iii Calculate the force required from the biceps muscle to hold the football in this position. *(3 marks)*

**Student Book
pages 44–45**

Key points

- 'Fitness for purpose' implies 'satisfactory for the task intended'.
- The British Standards Institute (BSI) and European Committee for Standardisation (CEN) set and test product standards.
- Density = mass ÷ volume (units: g/cm³).
- Good electrical conductors have a low resistance to the flow of electrical current.

CE (*conformité européene*) logo

3.1 Introduction to materials science

- The **BSI** and **CEN** set manufacturing standards so products are '**fit for purpose**'. The '**CE**' Mark shows a product has been quality tested and is safe to use.

> **1** *What do (a) BSI and (b) CEN stand for?*

The use a material is put to depends on its properties, and these properties are affected by its internal structure.
Scientists calculate a material's **density** in g/cm³ using the formula:

$$\textbf{density = mass} \div \textbf{volume}$$

- Carbon-fibre composites and aluminium and titanium alloys have a low density, making them suitable for lightweight bicycle frames that are easy to accelerate.

> **2** *A block of titanium has dimensions $1\,cm \times 2\,cm \times 5\,cm$ and a mass of $45\,g$. What is the density of titanium?*

- **Electrical conductivity** is found by measuring its resistance in ohms (Ω) using a multimeter. Copper is a good electrical conductor with a low resistance.

Key words: BSI, CEN, fit for purpose, CE, density, electrical conductivity

 Maths skills

Calculate the density of a material in two steps:
1. First calculate its volume using a formula, e.g. for a cube:
$$\text{volume = length} \times \text{width} \times \text{height (is cm}^3).$$
2. Then calculate its density using:
$$\text{density = mass} \div \text{volume}$$
Remember to add the correct unit g/cm³ after your answer.

**Student Book
pages 46–47**

3.2 Forces on materials

- **Hooke's law** applies to springs or wires (before its **elastic limit** is reached) when: Force (in N) = constant (in N/cm) × extension (in cm)
- The **spring constant** equals the **gradient** of the line on a force-extension graph.
- A wire, such as steel cable, with a high tensile strength resists stretching and needs a large force to break it. (Materials with high compressive strength resist crushing or squashing.)
- Scientists often measure the **tensile breaking strength** of a wire **not** in the unit of force (N), but in the unit of **stress**.
  Stress (N/mm²) = force (N) ÷ cross-sectional area (mm²)
  A wire's cross-sectional area = $\pi \times r^2$ (where r is its radius in mm)

> **1** *An aluminium alloy wire with a diameter of $2\,mm$ breaks when the tension is $1000\,N$. (a) What is its cross-sectional area? (b) Show its tensile breaking strength is approximately $320\,N/mm^2$.*

- **Brittle** materials crack easily. **Tough** materials absorb energy before breaking.

Key points

- Tensions are pulls that stretch, compressions are pushes that squash.
- Hooke's law: Extension is directly proportional to applied force.
- Stress = force ÷ area (the unit of stress is: N/mm²)

 Maths skills

If you know the diameter of a wire, halve this to calculate its radius. When using area = $\pi \times r^2$, square the radius then multiply it by π.

Key words: Hooke's law, elastic limit, spring constant, gradient, tensile breaking strength, stress, brittle, tough

Student Book
pages 48–49

3.3 Metals and alloys

Key points

- An alloy is a mixture of two or more elements, at least one being a metal.
- Metals and their alloys can be strong, malleable and good electrical and thermal conductors.
- Metals make excellent materials for structures.

- **Hard** materials are difficult to dent or scratch. Stainless steel surgical instruments are **strong** and hard.
- **Malleable** materials can be hammered into shape and rolled into sheets.
- **Stiff** materials are not **flexible**. The strength and stiffness of steel makes it ideal for supporting structures, but most types of steel suffer from rusting.

⫸ **1 Give an example of a steel structure.**

- The outer **electrons** in metals can flow from atom to atom, making metals good electrical and thermal conductors.
- A metal has an arrangement of positive **ions** 'in a sea' of negative electrons. The ions and electrons attract each other strongly. It is hard to break the bonds that hold the metal ions in position in its structure.

⫸ **2 Why are metals strong and hard?**

Key words: hard, strong, malleable, stiff, flexible, electron, ion

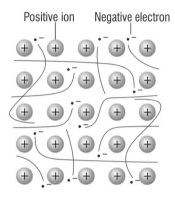

Positive ion   Negative electron

Metallic bonding – an arrangement of positive ions in a sea of negative electrons

AQA Examiner's tip

It doesn't matter if you've never seen a particular piece of equipment referred to in a question. Use your knowledge of material properties and common sense to work out the answer. You will sometimes be given some data to help you work out the correct answer.

e.g. Aluminium–titanium alloy is as strong as steel, but much less dense, making it far more suitable for aircraft frames. This is because light alloy aircraft need less fuel and power to lift them off the ground and fly.

Student Book
pages 50–51

3.4 Composites, wood and ceramics

Key points

- Composites contain a mixture of two materials. Composites benefit from the combined properties of their components.
- Wood is a strong natural composite, with a low density.
- Ceramics are non-metallic materials such as pottery and glass.

- **Composite** materials must complement each other. For example, 'glass-fibre reinforced plastic' uses strong, but brittle, glass with tough, but weak, plastic. The result is a strong and tough material, ideal for boat hulls, rather than a brittle and weak material.
- **Ceramics** are hard and brittle, have a high melting point and resist chemical attack. Racing cars have expensive carbon fibre–ceramic disc brakes instead of steel discs, which would bend at the high braking temperatures. By mixing hard, abrasion-resistant silicon carbide with strong, tough carbon fibres it lasts four times longer than conventional steel disc brakes. The light weight of carbon–ceramic composite also helps the car accelerate faster.

⫸ **1 Explain why the properties of carbon–ceramic composites are advantageous for the disc brakes of sports cars.**

Key words: composite, ceramic

AQA Examiner's tip

State your material, its useful property and why this property is advantageous. *Material, property, advantage* will often gain 3 marks!

Student Book
pages 52–53

3.5 Natural or synthetic?

- **Natural** fibres, such as cotton, silk, wool and leather, come from living things.
- **Synthetic** fibres, such as polyester and polyurethane **Lycra**, are man-made (from crude oil).
- A **polyester** cotton blend is an almost ideal clothing fabric. The cotton absorbs moisture, which makes for comfort, while the polyester is harder wearing and more crease-resistant than if the garment was made from cotton alone.

> **1** *Tight-fitting Lycra body suits and medical support socks encourage blood flow round muscles. How does (a) the flexibility and (b) the elasticity of Lycra make it suitable for these applications?*

- **Kevlar** is a lightweight, tough synthetic fibre, five times stronger than steel.

> **2** *What property of Kevlar makes it most suitable for (a) the lining of bicycle tyres and bullet-proof vests, (b) replacing damaged ligaments?*

Key words: natural, synthetic

Student Book
pages 54–55

3.6 Polymers

- Polymers are organic compounds with long-chain molecules made of 'many (similar) bits' called monomers.
- Many polymers are flexible and have low densities and low thermal conductivities. Wool and cotton are natural polymers. Plastics such as polyvinyl chloride (PVC – used for cling-film), polyester (used for clothing) and nylon (used for rope and fishing line) are synthetic polymers.
- There are strong covalent bonds between the atoms in each chain and weaker forces of attraction between the separate chains.

Scientists add chemicals to polymers to create intermolecular bonds, called **cross-links**, between the chains. The degree of cross-linking (or **branching** in the chains) determines a polymer's **strength**, **density** and **melting point**. Irregular branching along a chain makes it more difficult for the polymer chains to pack together, weakening the forces between chains. This decreases the polymer's density and lowers its melting point. Cross-links between chains strengthen the forces between chains. These increase the density and make it harder to melt the polymer.

A **thermoplastic polymer**, such as polyethene, used for carrier bags and plastic bottles, has few cross-links. Its relatively weak intermolecular bonds allow it to deform. A thermoplastic polymer is flexible and softens when heated, so is easy to mould and shape.

> **1** *Why do thermoplastic polymers soften when heated?*

A **thermosetting polymer**, such as vulcanised rubber for car tyres and shoe soles, has strong cross-links between its polymer chains. These cross-links make it rigid once set, and cause it to decompose, rather than melt, when heated.

Key words: cross-links, thermoplastic polymer, thermosetting polymer

**Student Book
pages 56–57**

3.7 Materials for sports, medicine and transport

Key points

- Insulation, density, strength, smoothness, flexibility and shock absorbency affect the suitability of a material for a product.
- Using synthetic polymers and composite materials has influenced design. People are no longer reliant on natural materials and pure metals.

- This table shows some examples of materials and their uses.

Type of material	Example of materials	Useful properties	For sports	For medicine	For transport
Metals	Aluminium with silicon or titanium alloy	Low density, strong	Tennis rackets	Replacement hip joints	Aircraft structures, bicycles
	Stainless steel	Strong, hard	Golf clubs	Surgical instruments/ implants	Ships tankers, car bodies/ exhausts
Polymers	Lycra	Flexible, elastic, low density	Body suits	Support stockings	Car bumpers
	Polypropene (PP)	Tough, flexible, low density	Floating rope	Sutures, valves, joints	Expanded for model aircraft
Ceramics	Ceramics mixtures	Hard, low thermal conductivity, resist chemical attack, high melting point	Clay pigeon shooting	Dentures	Space shuttle shell tiles
	Carbon-fibre ceramic		Racing car brakes		Catalytic converters
Composites	Titanium with carbon/ Kevlar fibres in resin	Low density, strong, tough	Tennis rackets, Golf clubs	Prosthetics	Bicycle frames

> **1** Choose separate materials for (a) sports, (b) medicine and (c) transport uses, and explain why the properties of the material are useful.

**Student Book
pages 58–59**

3.8 Standard procedures for testing materials

Key point

- Scientists use standard procedures to test the properties of materials.

- A **tough** material is not **brittle**, but absorbs energy before fracture.

> **1** If you stretch a material and it does not crack, but 'necks' before breaking, is it tough or brittle?

The flexibility /stiffness test

- A **hard** material does not dent or scratch. Test by dropping a mass onto a ball bearing on the sample and measuring the dent size.
- A **stiff** material is not **flexible**. Test by measuring the deflection of the end of a loaded rod.
- The unit of **stress** (N/mm²) is often used for compressive breaking strength and tensile breaking strength.

> **2** A compressive force of 540 kN distorts an aluminium rod of diameter 3 mm. (a) Use $A = \pi r^2$ to calculate its area of cross-section in mm². (b) Use stress = force ÷ area to calculate its compressive breaking strength in kN/mm².

AQA Examiner's tip

For a question on materials state:
Application: e.g. tennis racket
Material: e.g. titanium and graphite fibres in plastic
Properties: e.g. strong and lightweight
Advantage: e.g. lightweight rackets are more manoeuvrable.

1 When does a metal spring <u>not</u> obey Hooke's law?

2 What is the difference between *tensile* and *compressive* breaking strength?

3 What is the opposite of

   a flexible
   b brittle?

4 Suggest a use for titanium, based on its density being similar to that of bone.

5 What advantages has aluminium-titanium alloy over steel?

6 What properties of ceramics make them suitable for catalytic converters in car exhausts?

7 When placed in a measuring cylinder, a 27 g sample of aluminium displaces 10 cm^3 of water. What is the density of aluminium?

8 What property has the unit:

   a g/cm^3
   b Ω
   c N/cm
   d N/mm^2?

9 The strength and elasticity of rubber is improved by adding sulfur.

   a How does this create a thermosetting polymer?
   b Describe why the product is useful. [H]

Chapter checklist			✓ ✓ ✓
Tick when you have:			Metals and alloys ▢ ▢ ▢
reviewed it after your lesson	☑ ▢ ▢		Composites, wood and ceramics ▢ ▢ ▢
revised once – some questions right	☑ ☑ ▢		
revised twice – all questions right	☑ ☑ ☑		Natural or synthetic? ▢ ▢ ▢
Move on to another topic when you have all three ticks			Polymers ▢ ▢ ▢
	✓ ✓ ✓		Materials for sports, medicine and transport ▢ ▢ ▢
Introduction to materials science ▢ ▢ ▢			Standard procedures for testing materials ▢ ▢ ▢
Forces on materials ▢ ▢ ▢			

1 When designing sporting equipment, developers must ensure that their products comply with European safety laws. Products which meet these standards are labelled with the CE mark.

  a What is meant by 'fit for purpose'? *(1 mark)*

  b Select the **two** organisations which set and test standards for new products from the list below. *(2 marks)*

  BSI   ADH   BSM   CEN   JET

  c A new design of shot putt has been proposed. It is to be made of a new composite material, which has a density of 8.02 g/cm³. A materials scientist has been asked to test a shot putt made of this new material, to ensure that it meets expected standards.

   i The volume of the shot putt can be calculated using the equation:
   volume = 4/3 × 3.14 × radius³.
   Calculate the volume of a shot putt of diameter 12 cm. *(3 marks)*

   ii The Olympic standard for a men's shot putt is 7.26 kg.

   Mass = Density × Volume

   Use the equation above to determine the mass of the shot putt. Does the shot putt meet the Olympic standard?

   To gain full credit, you should show clearly how you arrived at your answer. *(3 marks).*

2 Polymers are used in a wide range of applications, from drinking cups to car tyres.

  a What is a polymer? *(1 mark)*

  b Draw a line to link each polymer to its properties, and an example of its use:

POLYMER	PROPERTIES	EXAMPLE OF USE
POLYETHENE	Strong, low density, flexible	Food containers
PVC	Transparent, hard, tough	Unbreakable 'glass' for watches
PERSPEX	Durable, tough, electrical insulator	Carrier bags
PET	Transparent, flexible, low density	Insulation for electrical cables

  (3 marks)

  c Describe the difference between a thermosetting and a thermoplastic polymer. [H] *(2 marks)*

3 Composites are used in the production of many different new products.

  a What is meant by a composite material? *(1 mark)*

  b *In this question you will be assessed on using good English, organising information clearly and using specialist terms where appropriate.*

  Name **two** products that are manufactured from composite materials. For each product give the properties of the composite material and explain why it is the best material for making that product. *(6 marks)*

4.1 Introduction to food science

Key points

- The FSA promote healthy eating, and make sure the food we eat is safe and labelled correctly.
- Defra ensure farmers produce healthy and sustainable food.
- Food scientists use a range of standard procedures to check nutrient levels and for dangerous microorganisms in food products.
- Food scientists include nutritionists, dieticians, food analysts, agricultural scientists and public health inspectors.

AQA Examiner's tip

Make sure that you know what the acronyms FSA (Food Standards Agency) and Defra (Department for Environment, Farming and Rural Affairs) stand for.

- The FSA (Food Standards Agency) and Defra (Department for Enviornment, Farming and Rural Affairs) are regulatory authorities. They are responsible for the safe production of food.
- **Food scientists** (including dieticians, public health inspectors and food analysts) work within the FSA to promote the importance of a healthy diet. They also ensure food is safe, and labelled correctly.
- **Agricultural scientists** work within Defra. They develop techniques to help farmers improve the quality and quantity of farm products. This ensures our food supply is healthy and sustainable.

▶ 1 *Name two regulatory authorities responsible for the food we eat.*

Food and agricultural scientists work in a range of different employments. Their roles include:

- Analysing the quality of food products.
- Checking food safety, and applying scientific techniques to keep food fresh, safe and attractive.
- Researching new food products, e.g. genetically modified crops.
- Researching methods for producing food more quickly and more cheaply, as well as conserving water and soil.
- Preparing diets for people in hospitals and in sport to help athletes improve their performance.

▶ 2 *Name three key roles a food scientist performs.*

Key words: food scientist, agricultural scientist

4.2 Food poisoning

Key Points

- Food poisoning is caused by the growth of microorganisms (usually bacteria) in food.
- For optimum growth bacteria need warmth, moisture and a food source.
- Common signs of food poisoning are stomach pains, diarrhoea, vomiting and fever.

Food poisoning is caused by the growth of microorganisms in food. The most serious types of food poisoning are caused by bacteria, and the toxins they produce. There are three main groups of bacteria that cause food poisoning:

- *Campylobacter* – found in raw meat, unpasteurised milk, and untreated water.
- *Salmonella* – found in raw meat, eggs, raw unwashed vegetables and unpasteurised milk.
- *E. coli* – some strains are beneficial, others cause illness. *E. coli* strains may be found in raw and undercooked meats, unpasteurised milk and dairy products.

To grow and reproduce effectively most bacteria need warmth, moisture and a food source.

▶ 1 *Name the three main groups of bacteria responsible for food poisoning.*

▶ 2 *What conditions do bacteria need to multiply?*

- Food poisoning symptoms include stomach pains, diarrhoea and vomiting. Most people require no treatment.

Key words: food poisoning

4.3 Food hygiene

Key points

- Food poisoning is avoided in commercial food preparation areas by ensuring work areas are clean, using sterile packaging, and by effectively controlling pests.
- Good personal hygiene is essential for employees in the food industry.
- Bacterial growth in food can be inhibited in a number of ways. These include: freezing, ultra-heat treatment, thorough cooking, pickling and drying.

Microorganisms can enter a food product from when the animal or crop is growing, until the food is eaten. If these microorganisms survive and multiply, they can cause illness. To reduce the risk of contamination people use:

- detergents, to ensure food preparation areas remain clean
- disinfectants, to kill bacteria on work surfaces
- sterile packaging materials and equipment – achieved using high temperatures or gamma rays
- appropriate waste disposal containers, removing waste from food preparation areas
- adequate control for insects, mice and other pests.

> **1** *List three techniques which can be used to reduce the risk of contamination.*

- Good personal hygiene and wearing protective clothing prevents potentially harmful microorganisms being transferred to foods.

> **2** *State three examples of protective clothing a person could wear when preparing food.*

- Bacterial growth can be slowed or stopped by changing the temperature. Bacteria multiply rapidly between about 5 °C and 65 °C. Most are killed at temperatures above 70 °C.
- **Food preservation** involves controlling bacteria by refrigeration, freezing, heating, cooking, drying, salting and pickling.

Technique to preserve food	How most bacteria are prevented from growing	Example food product
Refrigeration	Slows, but does not stop, the growth of bacteria.	Fruit juice
Freezing	Stops bacteria multiplying, but does not kill them	Ice cream
Heating – for example, ultra-heat treatment (UHT), where foods are heated to 132 °C for one minute	Kills virtually all microorganisms and their spores	Milk
Cooking	At the correct temperature, kills microorganisms	Roast chicken
Drying	Removes water, so bacteria cannot digest and absorb the food source	Rice
Salting	Bacteria lose water from their cells by osmosis. Most die, others are inactivated (can't reproduce).	Corned beef
Pickling	Acid (usually vinegar) is added to lower the pH of the food stuff. This inactivates most microorganisms	Pickled onions

> **3** *Name three food preservation techniques*

AQA Examiner's tip

Remember that **food hygiene** is important at all stages in food preparation – in the care, preparation and storage of food. Strict procedures are put into place to prevent food poisoning. In an exam answer, ensure you state:

- **how** good hygiene is maintained at each stage
- **why** this helps to prevent food poisoning.

Key words: food preservation, food hygiene

4.4 Standard procedures – microbiological techniques

Student Book
pages 72–73

Key points

- Microbiologists work in aseptic (sterile) conditions to ensure samples they are investigating are not contaminated.
- Microbiologists isolate individual colonies of bacteria using streak plates.
- The number of bacteria present in a sample can be calculated by carrying out a serial dilution.

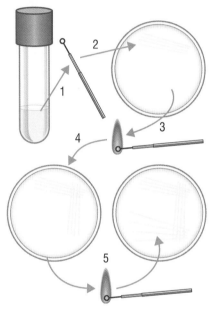

How to make a streak plate

If you are asked to describe how to make a streak plate, it may be easier to draw a series of labelled diagrams, rather than write a paragraph of text.

- Scientists use **aseptic** techniques to prevent unwanted microorganisms from entering a sample.

Standard procedure to sample the environment – to detect the presence of microorganisms:
- Wipe a sterile, cotton swab across the surface being sampled, then lightly across a sterile agar plate.
- Incubate for 48 hours to allow any microorganisms present to grow, then identify any organisms present.

➤ **1** *Why does the agar plate need to be incubated for a few days?*

Standard procedure to make **streak plates** – to isolate individual bacterial colonies:
1. Dip a sterilised wire loop into sample of bacteria.
2. Make four/five streaks across an agar plate.
3. Flame and cool the loop.
4. A second streak crossing the first spreads the bacteria.
5. Repeat steps 3 and 4 two more times.
6. Fix the lid (do not seal all the way round) and incubate upside-down.

Bacteria can be recognised by the colony they form. These differ in shape, colour, size and elevation.

Carrying out a **serial dilution** – allows individual bacterial colonies to be seen:
- By counting the number of colonies, the number of bacteria originally present can be calculated:
  **Number of bacteria (per cm³) in original sample =
  number of colonies × dilution of sample**
- If more than one type of colony is visible, more than one species of bacteria is present. Apply the above equation to each species separately.

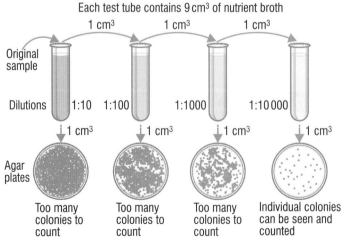

Carrying out a serial dilution

➤ **2** *Why do scientists not count individual bacteria?*

Key words: aseptic, streak plate, serial dilution

Student Book
pages 74–75

4.5 Bread, beer and wine production

Key points

- Fermentation produces ethanol (alcohol) and carbon dioxide from glucose (sugar).
- Fermentation is used to make bread, beer and wine

- **Yeast** is a type of fungus. It is needed to make bread, beer and wine. These three products are made using a chemical reaction called **fermentation**.
- Fermentation is an example of anaerobic respiration. The yeast respires without oxygen to ferment sugar, producing alcohol and carbon dioxide. Fermentation can be summarised in this equation:

glucose $\longrightarrow$ ethanol (alcohol) + carbon dioxide

$$C_6H_{12}O_6 \longrightarrow 2C_2H_5OH + 2CO_2$$

- Enzymes in yeast speed up the process of fermentation. The ideal conditions for fermentation are a good supply of glucose, with no oxygen present, and at a temperature between 15°C and 25°C.

1 *What are the optimum conditions for fermentation?*

- Bread is made by mixing flour, water, sugar and yeast. The yeast ferments the sugar, producing carbon dioxide which makes the bread dough rise. When the dough is baked the ethanol evaporates.
- Beer is made using hops, barley and yeast. The yeast ferments the sugar (maltose) in the barley, to produce alcohol.
- Wine is made from grapes and yeast. The yeast ferments the grape sugars, to produce alcohol.

2 *Why does bread not contain alcohol?*

Key words: yeast, fermentation

AQA Examiner's tip

Note carefully how the examiner wants you to write an answer. If a question asks you to 'state the word equation for fermentation', write your answer in words, not as chemical formulae. For example, you would write 'carbon dioxide', not 'CO_2'.

Student Book
pages 76–77

4.6 Cheese and yoghurt production

Key points

- Cheese is made by adding bacteria and rennet to milk.
- Yoghurt is made by adding bacteria to boiled milk. The milk is kept warm, allowing the bacteria to ferment the lactose (milk sugar).

- Microorganisms are used in food production because they reproduce rapidly, are easy to manipulate, and are reliable.
- In cheese and yoghurt production, bacteria **ferment** sugars found in milk. Lactose (milk sugar) is converted into lactic acid. This acid gives these products their characteristic tangy taste.
- Cheese is made from the milk of many animals. Bacteria and rennet are added to curdle the milk. The whey (liquid) is drained off, and the curds (solids) pressed to make cheese.

1 *Which two substances have to be added to milk to turn it into cheese?*

- Yoghurt is made by adding bacteria to pasteurised milk (heated to kill harmful bacteria). The milk is then kept warm for several hours. During this time the bacteria multiply, and ferment lactose into lactic acid. The lactic acid curdles the milk into yoghurt and hinders the growth of harmful bacteria in the yoghurt.

2 *Why is lactic acid useful in yoghurt production?*

Key words: ferment

Bump up your grade

Cheese and yoghurt are produced from milk, using similar techniques. Learn the similarities and differences between the two production methods; this will enable you to explain how these lead to different food products.

Student Book
pages 78–79

4.7 Growing crops

- Plants make their own food, through **photosynthesis**. It is summarised in the equation below:

$$\text{carbon dioxide} + \text{water} \xrightarrow[\text{in chlorophyll}]{\text{light energy trapped}} \text{glucose} + \text{oxygen}$$

$$6CO_2 + 6H_2O \longrightarrow C_6H_{12}O_6 + 6O_2$$

- The rate of photosynthesis is affected by light intensity, water availability, carbon dioxide concentration and the surrounding temperature.
- Commercial plant-growers use controlled environments to ensure a high rate of photosynthesis.

There are two main types of farming:

Intensive farming is where the growth of crops and animals are maximised, often through the use of chemicals, resulting in high yields.

- Common chemicals used include fertilisers (add nutrients), pesticides (kill pests), herbicides (destroy weeds) and insecticides (kill insects).
- Intensive farming techniques are less labour intensive, allowing food to be produced more economically.

Organic farming is a natural method of producing crops and rearing animals. For example, weeds are removed by hand or machine and manure is added to improve soil quality.

- Biological control techniques are used to control pests. Predators (such as parasitic wasps or ladybirds) are bred in large numbers. They are released onto crops, where they kill the pests.
- Yields are generally smaller, resulting in more expensive products. However, many people believe organic food is healthier and tastes better.

▶ **1** *Name an advantage of (a) intensive farming and (b) organic farming.*

▶ **2** *What is meant by biological control?*

Key words: photosynthesis, intensive farming, organic farming

Key points

- The rate of photosynthesis is affected by light intensity, water availability, carbon dioxide concentration and temperature.
- Intensive farming uses chemicals to produce crops as efficiently as possible.
- Organic farmers add nutrients to soils using manure and compost. They kill pests using biological control, and remove weeds by hand or using a machine.

AQA *Examiner's tip*

Make sure you know the main differences between the two types of farming – organic and intensive. Try producing a summary table to help remember the differences (as in 4.9).

Student Book
pages 80–81

4.8 The use of chemicals in intensive farming

- The reaction between an acid and an alkali is known as a **neutralisation reaction**. Industrial chemists use neutralisation reactions to manufacture fertilisers. E.g.

$$\text{nitric acid (acid)} + \text{ammonia (alkali)} \longrightarrow \text{ammonium nitrate}$$

$$HNO_3 \text{ (aq)} + NH_3 \text{ (aq)} \longrightarrow NH_4NO_3 \text{ (aq)}$$

Agricultural chemicals are produced as efficiently as possible to ensure raw materials are not wasted, and costs are kept low. To increase efficiency, industrial chemists:

- build chemical plants near to raw materials to reduce transport costs
- minimise and recycle any chemical by-products
- ensure a good rate of chemical reaction
- maximise the yield from a reaction.

▶ **1** *State four ways chemical processes can be made more efficient.*

Key points

- Neutralisation reactions occur when acidic and alkaline solutions react together.
- Fertilisers can be made using the neutralisation reaction between ammonia solution, and either nitric acid or sulfuric acid.

Key words: neutralisation reaction

4.9 Rearing animals

Key points

- Intensively farmed animals are kept in a strictly controlled environment. This makes the animals increase in size quickly.

- Factors that are controlled include their diet, size and temperature of enclosure and the routine use of drugs.

- Organically farmed animals are free to roam in large enclosures and are only given drugs when ill.

Battery farmed hens are kept in a carefully controlled environment

Organically farmed hens will produce fewer eggs than battery farmed hens but have a higher life expectancy

- Intensively farmed animals are kept in strictly **controlled environments**. This makes their meat and products cheaper, as the animals increase in size quickly. Animals farmed organically need more space, more time and more labour to look after them. This means costs are higher.

- The main differences between intensive and organic methods of animal production are shown in the table:

Factors that can be controlled	Intensively reared animals	Organically reared animals
Food supply	Animals fed a high protein diet, to rapidly increase their body mass.	Animals fed organic food.
Temperature	Animals are kept indoors, in a warm environment. Animals waste less energy heating their own bodies.	Animals would normally live outdoors in the day time. At night or in bad weather, animals may be kept inside.
Space	Restricted movement. Animals do not waste energy moving around.	Animals are allowed to roam as freely as possible.
Use of drugs	Antibiotics are given regularly to prevent the spread of disease.	Antibiotics are not used, unless an animal is ill.
Enclosure safety	Animals are kept safe from predators.	Increased risk from predators. May be kept indoors at night to minimise risk.
Infection control	Animals are close-packed, so infections could spread quickly. Animals are checked regularly. Enclosures are under strict quarantine.	Animals kept outside are more likely to catch infections from wild animals.
Waste efficiency	Animal waste can be collected and converted into biogas.	Animal waste is not available for biogas.

▶ 1 *Name an advantage and disadvantage of rearing animals intensively.*

▶ 2 *Name an advantage and disadvantage of rearing animals organically.*

AQA *Examiner's tip*

When asked questions on how animals are reared using the two types of farming, make sure you stick to the facts. For example, a question may ask: 'Give one reason why some people do not agree with intensive farming'. Answers such as 'Intensive farming is cruel' will not gain credit. To gain a mark you could say: 'Animals are kept in small spaces in intensive farming, which some people think is cruel'.

Key words: controlled environment

Student Book
pages 84–85

4.10 The impact of intensive farming on the environment

Key points

- Intensive farming can pollute rivers and lakes, cause soil erosion, and result in the removal of nutrients from soil.
- Chemicals used in intensive farming can kill untargeted animals. Animals may also lose their habitats when hedgerows are removed.

This barren hillside has been severely affected by soil erosion, as a result of over-grazing by farm animals

- **River pollution** – Excess fertiliser dissolves in rainwater, and is washed into lakes and rivers. This causes rapid algal growth, covering the water surface. Light no longer reaches the lower plants. These plants die and are broken down by bacteria. The decaying process uses up lots of oxygen in the water, making it difficult for animals to survive; many die. This process is known as **eutrophication**.
- **Soil erosion** – Over-grazing destroys vegetation whose roots bind soil together. Barren soil dries out easily, increasing the risk of wind or water erosion. Hedgerow removal to create larger fields destroys habitats and causes soil erosion; hedges act as wind-breaks, preventing soil being blown or washed away.

> **1** *Which natural processes erode soil?*

- **Poisoning wildlife** – Chemicals can accumulate in food chains and result in large numbers of wildlife being killed. They can be passed along a food chain until they reach fatal levels – killing many top predators.
- **Monoculture** means always growing one crop on an area of land. This can result in:
  - The same nutrients constantly being removed from the soil, requiring a greater use of fertilisers.
  - Crop pests being concentrated in one area, leading to greater use of pesticides.
  - Reduction of **biodiversity**, as the habitat for some animals is removed as only one plant is grown.

> **2** *What does monoculture mean?*

Key words: eutrophication, monoculture, biodiversity

Student Book
pages 86–87

4.11 Investigating plant growth

Key points

- Plant growth can be measured in a number of ways including height and mass of plant, or size and number of leaves.
- Plants grown hydroponically are grown in a solution of water containing dissolved minerals.
- Some of the main factors affecting plant growth are light, water, space and availability of nutrients.

- Plants can be grown in water which contains dissolved minerals. This is known as **hydroponics**. It is often used by commercial growers, because it enables plants to grow quickly. It also enables agricultural scientists to study plant growth.

Factors affecting plant growth:

- **Light availability** – an intense light source ensures a plant photosynthesises rapidly. Artificial lights can be used to produce 'summer conditions' all year around, increasing yields.
- **Water availability** – plants need an adequate water supply to photosynthesise rapidly.
- **Space** – plants need enough space to absorb nutrients and water from the soil and gather sunlight. If plants become overcrowded, they tend to be smaller and produce lower yields.
- **Nutrient availability** – plants need minerals for healthy growth. Farmers need to recognise mineral deficiency symptoms, so that they can add appropriate chemicals to the soil (normally fertilisers).

Mineral	Role of mineral	Deficiency symptoms
Nitrates (contain nitrogen)	Needed to make DNA and amino acids. The amino acids join to form proteins needed for healthy leaf growth.	Older leaves are yellowed, growth is stunted.
Phosphates (contain phosphorus)	Needed for healthy roots.	Younger leaves have a purple tinge and poor root growth.
Potassium	Needed for healthy leaves, flowers and high fruit yield.	Yellow leaves, with dead areas on them.
Magnesium	Needed for making chlorophyll molecules.	Pale or yellow leaves.

AQA *Examiner's tip*

To ensure scientists are studying the effect of only one variable, a control should be used. This is a second plant which is exposed to the same conditions as the plant being tested, except for the factor under investigation.

▻ **1** *How can you tell if a plant is lacking in water?*

▻ **2** *How could you tell a plant was lacking in magnesium?*

Key words: hydroponics, nitrate, phosphate, potassium, magnesium

---

Student Book pages 88–89

Key Points

- When particles of reactants collide together with sufficient energy, they react to produce a new product. This is known as 'collision theory'.
- The rate of a chemical reaction can be affected by: concentration, temperature, surface area and the use of a catalyst.
- Chemical reactions can be speeded up by: increasing the concentration of a solution, increasing the temperature, increasing the surface area of a solid, using a catalyst.

AQA *Examiner's tip*

Do not confuse the size of pieces of a solid, and its surface area. For example, if a large piece of solid material is broken into many pieces, the combined surface area of these smaller pieces is much larger than that of the large piece of solid.

4.12 Rates of reaction

- Chemical manufacturers need to be able to produce **products** efficiently and economically. They need to understand how different factors affect the rate of chemical reactions.
- Substances that react together are called **reactants**. Their particles may be atoms, ions or molecules.
- In order for a chemical reaction to occur, the particles must (i) collide, and (ii) collide with enough energy to cause a reaction.
- The minimum energy required for a reaction to occur is known as the **activation energy**.

Several factors affect the rate of a chemical reaction. These include:

- **Concentration** – An increase in concentration increases the number of collisions in a given time, and so increases the rate of reaction.
- **Temperature** – An increase in temperature increases both the frequency of collisions and the energy of collisions, as particles move faster. This increases the rate of reaction.
- **Surface area** – An increase in surface area of a solid reactant increases the frequency of collisions. This also increases the rate of reaction.
- **Use of catalysts** – Adding a catalyst increases the rate of reaction, as catalysts reduce the activation energy required in a chemical reaction.

▻ **1** *State the four factors which can affect the rate of a chemical reaction.*

Key words: products, reactants, activation energy

4.13 Chemical yields

Key points

- The amount of product collected from a chemical reaction is known as the yield.
- The theoretical yield is the maximum amount of product a reaction could produce.
- The actual yield is the amount of product which is collected from a reaction.
- Percentage yield is the proportion of a product collected compared to the theoretical yield, expressed as a percentage.

- When reactants combine, the amount of useful product made is called the **yield**. The yield a chemical reaction is expected to produce is known as the **theoretical yield**.
- In reality, theoretical yields are never reached. The amount of product you actually get is called the **actual yield**. It may be lower than the theoretical yield.

▐▶ **1** *What is the difference between theoretical and actual yield?*

- The **percentage yield** is a measure of the actual yield of a product, compared to the theoretical yield.

$$\text{Percentage yield} = \frac{\text{Actual yield}}{\text{Theoretical yield}} \times 100$$

Key words: yield, theoretical yield, actual yield, percentage yield

4.14 Reversible reactions

Key points

- Reversible reactions are chemical reactions in which the products can re-form the reactants. [H]
- Ammonia is manufactured using the Haber process. [H]
- For all reversible reactions: increasing temperature favours the endothermic reaction; increasing pressure favours the side of the reaction with fewest molecules of gas. [H]

Bump up your grade

The key to a good answer to a Haber process question is in explaining **why** a temperature of 450°C, high pressure and a catalyst are used. You should link this work with collision theory which explains rates of chemical reactions.

Some chemical reactions release energy – these are known as **exothermic** reactions. Others take in energy during the reaction – these are **endothermic** reactions.

In a reversible reaction, the products can react together to re-form the reactants. One of the reactions is exothermic, and the other is endothermic.

Ammonia is formed when gaseous hydrogen and nitrogen react together. Some of the product (ammonia) dissociates back into the reactants.

Word equation: **nitrogen + hydrogen ⇌ ammonia** (Exothermic →)

Symbol equation: $N_2(g) + 3H_2(g) \rightleftharpoons 2NH_3(g)$ (Endothermic ←)

In this reversible reaction, two factors affect the yield of the product. These are:
- Temperature: The higher the temperature, the lower the yield of ammonia.
- Pressure: The higher the pressure, the higher the yield of ammonia.

Ammonia is produced using the Haber process. In this process, a balance is found between a high yield, keeping a reasonable rate of reaction, and the costs of producing these reaction conditions. The chosen conditions are:
- 450°C (increases the rate of reaction, but reduces the yield)
- High pressure, 200 atmospheres (to increase the rate of reaction and yield)
- Use of an iron catalyst (to increase the rate of reaction)

▐▶ **1** *What are the conditions used in the Haber process?*

Key words: exothermic, endothermic

Student Book
pages 94–95

4.15 Selective breeding and genetic engineering

- Farmers choose plants or animals with the best characteristics to reproduce. For example, they pick sheep that produce lots of wool, or chickens that lay lots of eggs.
- Farmers then choose the best offspring, and mate them again. This technique, known as **selective breeding**, is continued over many generations.
- Advantages of selective breeding include higher yields and products available for longer periods of the year. It extends an organism's tolerance and can produce a more uniform crop making it easier to harvest.
- Disadvantages of selective breeding include a reduction in variation, reduction in the gene pool and as a result animals and plants may have poorer health.

▶ **1** *Give two advantages and two disadvantages of selective breeding.*

Producing animals with desired characteristics through selective breeding is slow and not very accurate. Scientists can now alter an organism's genes to produce desired characteristics – **genetic engineering**.

It can happen in one generation. Genes from a different organism (known as foreign genes) are put into plant or animal cells at an early stage in their development. As the organism develops it will display the characteristics of the foreign genes, as well as its own.

Examples of genetically modified (GM) products include: tomatoes (which are frost-resistant), cotton (with a high yield and pest resistance), and bacteria (which produce insulin for use by diabetics).

▶ **2** *Give two advantages of genetic engineering over selective breeding.*

Key words: selective breeding, genetic engineering.

Higher

Key points

- Farmers selectively breed plants and animals by breeding organisms with desired characteristics.
- Scientists are able to alter an organism's genes to produce organisms with desired characteristics – genetic engineering. [H]
- Genetic engineering is a faster and more accurate technique than selective breeding. [H]

AQA Examiner's tip

The genetic modification of plants and animals is a controversial issue. If you are asked to 'Discuss the ethical implications of genetic engineering', you should include arguments both for and against the process.

Student Book
pages 96–97

4.16 Standard procedures used in food science

- Food scientists perform a number of **standard procedures** to determine what is in food and drink products.
- Some of these detect the presence of a substance – this is called a **qualitative test**. It tells you whether a specific substance such as an additive or an unwanted microorganism is present. Making a streak plate to identify the type of bacterium present is an example of a qualitative test.
- To find out how much of this substance is present, a **quantitative test** must be used. Carrying out a serial dilution to make an accurate bacterial count is an example of a quantitative test.

▶ **1** *Which type of test would you perform if you wanted to find out how much fat was in a packet of crisps?*

Key words: standard procedure, qualitative test, quantitative test

Key Points

- Food scientists monitor the contents, quality and safety of food products, as well as developing new products.
- Qualitative tests identify the presence of a substance.
- Quantitative tests determine how much of a substance is present.

1 Microorganisms play an important role in the production of a number of food and drink products. To make bread, bakers add yeast to flour, water and sugar.

    a What process does the yeast carry out?

    b What are the two products of this reaction?

2 a What type of microorganism is the most common cause of food poisoning?

    b Name an example of a species of microorganism commonly associated with food poisoning.

    c Describe the symptoms of mild food poisoning.

3 a What is meant by the term monoculture?

    b Name two disadvantages of farming in this way.

4 Explain how a farmer can selectively breed chickens to be good egg-layers.

5 Describe the standard procedure which would be used to make a streak plate.

6 a What is meant by a reversible chemical reaction? [H]

    b Write a balanced symbol equation for the production of ammonia. [H]

    c Explain why these conditions are used. [H]

7 a What is genetic engineering? [H]

    b Give an advantage, with an example, of a genetically engineered crop. [H]

Chapter checklist

Tick when you:				Growing crops			
reviewed it after your lesson	✓	☐	☐	The use of chemicals in intensive farming	☐	☐	☐
revised once – some questions right	✓	✓	☐	Rearing animals	☐	☐	☐
revised twice – all questions right	✓	✓	✓	The impact of intensive farming on the environment	☐	☐	☐

Move on to another topic when you have all three ticks

				Investigating plant growth	☐	☐	☐
Introduction to food science	☐	☐	☐	Rates of reaction	☐	☐	☐
Food poisoning	☐	☐	☐	Chemical yields	☐	☐	☐
Food hygiene	☐	☐	☐	Reversible reactions	☐	☐	☐
Standard procedures – microbiological techniques	☐	☐	☐	Selective breeding and genetic engineering	☐	☐	☐
Bread, beer and wine production	☐	☐	☐	Standard procedures used in food science	☐	☐	☐
Cheese and yoghurt production	☐	☐	☐				

1 Microbiologists are scientists who specialise in working with microorganisms.

  a Suggest one role a microbiologist may carry out in the food industry.   *(1 mark)*

  b Copy and complete the following sentences, using the words below. Each word can
     be used once, more than once or not at all:

     bacteria   butter   yeast   bread   cheese   viruses

     Some microorganisms are used to produce food products. For example,
     .... are used to manufacture yoghurt and ....
     .... is used during the manufacture of .... and wine.
                                                                              (4 marks)

  c Name the useful reaction yeast performs during the manufacture of beer.   *(1 mark)*

  d Complete the word equation for the reaction in **1c**.   *(2 marks)*

     .... $\longrightarrow$ .... + ...

  e Describe simply how yoghurt is manufactured.   *(4 marks)*

2 Farmers need to make informed decisions over how to raise their livestock. Many make
  decisions based on economic considerations; others put ethical reasons first.

  a Describe what is meant by each of the following types of farming:

     i   Intensive farming   *(1 mark)*

     ii Organic farming.   *(1 mark)*

  b Copy the table below, and complete by using the statements beneath: *(4 marks)*

Factors that can be controlled	Intensively reared animals	Organically reared animals
Temperature		
Space		
Use of drugs		
Enclosure Safety		

  • Antibiotics are given to prevent the spread of disease.
  • Antibiotics only given if an animal is ill.
  • Animals are allowed to roam as freely as possible.
  • Movement restricted to ensure energy used for growth.

  • Animals normally outdoors. Animals use energy to heat their bodies.
  • Live outdoors, so higher risk from predators.
  • Animals are kept inside, safe from predators.
  • Animals are kept indoors in a warm environment

  c Organically reared meat is more expensive to produce than intensively farmed meat.

     i   State **one** reason why organic meat is usually more expensive to manufacture.
                                                                              (1 mark)

     ii Suggest and explain why some people are prepared to pay more for organically
         produced foods.   *(2 marks)*

  d *In this question you will be assessed on using good English, organising information
     clearly and using specialist terms where appropriate.*

     To be classified as 'organically reared', livestock must be fed an organic diet.
     State and explain two ways a farmer could manage to produce crops
     efficiently without using chemicals.   *(6 marks)*

AQA Examiner's tip

A 'Describe' question means what it says. Do not try to explain anything – just write what is happening.

AQA Examiner's tip

Questions are always labelled with the number of marks available for a response. For each mark you need to state one point. For example, part **2cii** is worth two marks, and therefore you have to make two separate points. When a question asks you to explain something try and link these points together.

Student Book pages 104–105

5.1 Introduction to analytical science

Key points

- Daily tasks of analytical scientists involve quality control, research and development, and identification of materials.
- Analytical scientists work in forensics, environmental protection, and healthcare and pharmaceuticals.

Examples of the work of analytical scientists include:

- **Quality control**, such as monitoring the production of foods, drinks, cosmetics and pesticides.
- **Research and development**, such as improving the quality of foodstuffs and developing new drugs.
- **Identification**, such as analysing body tissues and fluids to help diagnose diseases. It is also used for analysing materials found at crime scenes.

1 ***Give three examples of evidence that could be left at a crime scene.***

Two organisations which use analytical techniques are:

- **Defra** (the Department for Environment, Food and Rural Affairs) helps the government to make policies and write laws covering environmental issues.
- The **Health Protection Agency** responds to health hazards and emergencies by providing advice and information.

2 ***Why is the work of Defra important?***

Key words: quality control, Defra, Health Protection Agency

Student Book pages 106–107

5.2 Distinguishing different chemicals

Key points

- The properties of substances can be used to help identify the substances.
- Ionic compounds have high melting points and boiling points. This is because they need lots of energy to break the strong bonds between positive and negative ions in their giant structures.
- Most covalent compounds have low melting points and boiling points.

The closely packed ions in the giant structure of sodium chloride

- Ions are atoms that have gained or lost electrons. In **ionic compounds**, such as sodium chloride, NaCl, the strong electrostatic attraction (or bond) between the positive (metal) and negative (non-metal) ions gives the compound a giant, regular structure. It takes a lot of energy to break the many bonds and separate the ions, so ionic compounds have a high melting point.
- To write formulae of ionic compounds, match the charges so they cancel out (add up to zero). Sodium nitrate is $NaNO_3$, but magnesium nitrate is $Mg(NO_3)_2$.

1 ***What is the formula for (a) copper(II) sulfide, (b) copper(II) sulfate, (c) sodium carbonate, (d) lead nitrate?***

- **Covalent compounds** share electrons. Most are easy to melt and boil because, although the covalent bond between the atoms is strong, the forces between the individual molecules are weak.
- Organic compounds, such as ethanol C_2H_5OH and glucose $C_6H_{12}O_6$, come from living things and have covalent bonds within their molecules.

2 a ***What sort of bonding has lead bromide, $PbBr_2$ (ionic or covalent)?***

b ***Does ethanol have a high or a low melting point?***

Key words: ionic compound, covalent compound

AQA *Examiner's tip*

Practise writing the formulae for simple ionic compounds. Remember:
-ides e.g. sulfide, S, have just the element
-ates e.g. sulfate, SO_4, also contain oxygen.

The use of science in analysis and detection

5.3 Testing for ions

Student Book pages 108–109

Key points

- The colour of a precipitate can help you identify a substance.
- The sodium hydroxide test (NaOH) uses precipitation reactions to identify metal ions.

Bump up your grade

Learn how to use precipitation reactions (5.3) alongside using flame tests (5.5) to detect positive metal ions.

- A precipitate is an insoluble solid formed when mixing two solutions. The colour of a precipitate can help you identify a substance.
- The table shows the results of the **sodium hydroxide (NaOH) test** for some positive metal ions:

Metal ion in solution	Observation after adding NaOH solution
Lead (Pb^{2+})	White precipitate, which dissolves if you add more NaOH
Calcium (Ca^{2+}),	White precipitate, which doesn't re-dissolve if more NaOH is added
Copper(II) (Cu^{2+})	Blue, jelly-like precipitate
Iron(II) (Fe^{2+})	Green-grey, jelly-like precipitate
Iron(III) (Fe^{3+})	Red-brown, jelly-like precipitate

- e.g. Adding sodium hydroxide (NaOH) to river water, polluted with copper(II) sulfate, produces a blue **precipitate** (copper(II) hydroxide):

$$CuSO_4 + 2NaOH \longrightarrow Cu(OH)_2 + Na_2SO_4$$

1 *(a) What is the precipitate produced when calcium nitrate reacts with sodium hydroxide? (b) What metal ion does this reaction identify? (c) What is the formula for the precipitate?*

- Tests for negative non-metal ions:

Carbonate ($CO_3{}^{2-}$)	Add dilute acid	Carbon dioxide gas given off, which turns limewater cloudy
Chloride (Cl^-)	Add a few drops of dilute nitric acid, then a few drops of silver nitrate solution	White precipitate
Sulfate ($SO_4{}^{2-}$)	Add a few drops of dilute hydrochloric acid, then a few drops of barium chloride solution	White precipitate

2 *Is a precipitate (a) soluble or insoluble, (b) a solid or liquid?*

Key words: precipitate

5.4 Breathalysers

Student Book pages 110–111

Key point

- You test for ethanol with warm acidified potassium dichromate, which changes colour from orange to green.

- Breathalysers (i.e. *breath analysers*) test for **ethanol** (C_2H_5OH) in your breath.
- In 1954, Borkenstein invented a breathalyser using the ethanol + **acidified potassium dichromate** reaction to produce a colour change. Since the 1980s, in order to achieve greater accuracy and a faster response time, breathalysers now use either infrared spectroscopy or fuel cell technology.

1 *Why are (a) accuracy and (b) response time important in the design of breathalysers?*

AQA Examiner's tip

In the examination you may be asked to 'discuss the advances in breathalyser technology over time', but you do not need to remember all the details of the breathalyser design.

Alcohol test

5.5 Flame tests

The flame test standard procedure:

- Wearing eye protection, clean a nichrome wire loop in concentrated hydrochloric acid.
- Dip the loop into the sample, place in the edge of a hot Bunsen flame and record the colour.

Metal ion	Flame colour
Sodium Na$^+$	Bright yellow
Potassium K$^+$	Lilac
Copper Cu^{2+}	Green-blue
Calcium Ca^{2+}	Brick red

⟹ **1** *Hot corrosive hydrochloric acid could spit from the wire loop during a flame test. What precaution should you take?*

⟹ **2** *Why doesn't a flame test work on a mixture of salts?*

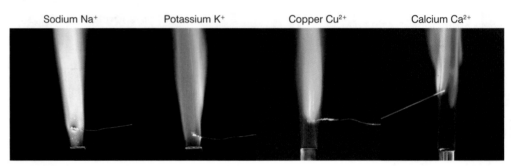

Sodium Na$^+$ Potassium K$^+$ Copper Cu^{2+} Calcium Ca^{2+}

Flame test colours

5.6 Balanced equations

- The **relative atomic mass** of hydrogen, H, is 1; and of oxygen, O, is 16.
  The **relative formula mass** of water, H_2O, is $(1 \times 2) + 16 = 18$
  18 g of water contain a **mole** of molecules.

⟹ **1** *Show that one mole of nitric acid, HNO_3, has a mass of 63 g. (The relative atomic mass of N = 14)*

- To write a balanced equation put numbers in front of the formulae to ensure each element has the same number of atoms on both sides of the equation.
  e.g. $H_2SO_4 + 2NaOH \longrightarrow Na_2SO_4 + 2H_2O$
  has 6 oxygen atoms on each side.

⟹ **2** *Two moles of nitric acid react with one mole of calcium carbonate. The products are calcium nitrate, water and a gas.*

  a *What is the gas?*

  b *Write a word equation for the reaction.*

  c *Write a balanced symbol equation for the reaction.* [H]

A mole of a substance is its relative formula mass in grams

Key words: relative atomic mass, relative formula mass, mole

Student Book
pages 116–117

5.7 Titrations

Key points

- In a titration you add solution from a burette to a pipette-measured volume of the other reactant. An indicator shows the end point.

- Taking care, repeat with clean apparatus until results are repeatable to improve accuracy.

Bump up your grade

To improve your mark on the Higher Tier paper, practise titration calculations until you are confident with the method.

The standard procedure for an acid–base **titration** to analyse the amount of substance in a sample:

- Wearing eye protection, **pipette** a measured volume of alkali into a conical flask and add **indicator**.
- Fill a **burette** with acid and slowly add the acid to the alkali, swirling the flask.
- Stop at the **end point** (when the indicator just changes colour) and record the volume of acid used.
- Repeat until two titration volumes agree (V in cm³).
- Calculate the moles of acid used to neutralise the alkali using:
  (V (in cm³) ÷ 1000) × concentration in moles/dm³.
- Calculate the mass of alkali in sample using:
  moles used × relative formula mass of alkai.

▶ 1 *10 cm³ of sodium hydroxide solution neutralise 25 cm³ of hydrochloric acid (of concentration 0.2 mol/dm³).*

  a *How many moles of HCl are present in the 25 cm³ of acid?*

  b *How many moles of NaOH neutralised this acid?*

  (NaOH + HCl $\longrightarrow$ NaCl + H_2O)

  c *What is the relative formula mass of NaOH?*
  (Relative atomic masses: H = 1, O = 16, Na = 23)

  d *What mass of NaOH neutralised the acid?*

Key words: titration, end point

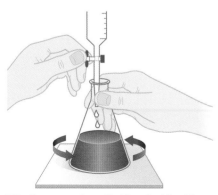

Titration procedure. Swirl and add acid drop by drop near the end point.

Student Book
pages 118–119

5.8 Chromatography

The Chromatography standard procedure:

- Draw a base line on the chromatography paper. Wearing eye protection, 'spot' the chromatography paper on the base line.
- Place paper in the solvent.
- Leave for the **solvent** to rise and mark where the solvent got to (the solvent front) before drying.
- Compare samples from the spots obtained by measuring and then by calculating the **retention factor (R_f)**:

R_f = distance travelled by substance ÷ distance travelled by solvent.

> ▶ **1** *A chemical moves 3 cm when the solvent travels 10 cm. What is its R_f?*

- In paper chromatography, different chemicals in the mixture separate at different rates. Some cling more strongly to the paper (**stationary phase**), some are more soluble in the solvent (**mobile phase**). When a mixture is insoluble in water scientists use a different solvent with thin layer chromatography.

> ▶ **2** *When is thin layer chromatography needed rather than paper chromatography?*

Key points

- Chromatography is used to separate and compare mixtures such as inks.
- In paper chromatography the stationary phase is the paper containing trapped water molecules.
- In thin layer chromatography (TLC) the stationary phase is a layer of powder, coated onto a sheet.
- Different substances travel different distances depending on how each is attracted to the stationary phase and the mobile phase or solvent. [H]

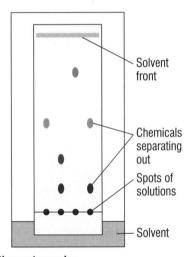

Chromatography

Solvent front

Chemicals separating out

Spots of solutions

Solvent

Bump up your grade

The key steps in the chromatography method:
1. Wearing eye protection, dissolve sample in solvent and place 'spot' on pencil line 2 cm from bottom of paper on the base line.
2. Add 1 cm depth of solvent in a beaker. Leave until solvent *almost* reaches top of the paper. Mark solvent front.
3. Take out and dry, then measure the distance travelled by the sample and calculate the retention factor (R_f).

Key words: solvent, retention factor (R_f), stationary phase, mobile phase

Student Book
pages 120–121

5.9 Microscopic evidence

- A **comparison microscope** is two microscopes side-by-side – useful to compare scratch marks on bullet cartridges.
- A **polarising microscope** brightens the object and darkens the background – useful to **contrast** paint and fibres.
- A **scanning electron microscope** (SEN) gives great **resolution** (sharpness and magnification) – useful for identifying **trace evidence** such as pollen grains and fibres. These each have different diameters, colours and surface patterns.

> **1** *Which microscope is useful to check if bullets were fired from the same gun?*
>
> **2** *Which microscope has the best resolution?*
>
> **3** *Which trace evidence could you use to identify a car involved in a hit-and-run accident?*

Key words: comparison microscope, polarising microscope, contrast, scanning electron microscope, resolution, trace evidence

Key points

- Forensic scientists use microscopes to compare features of different objects.
- Polarising and electron microscopes are used to improve contrast and resolution. These provide more precise and reliable evidence than simple light microscopes.
- The size, surface pattern and colour of pollen grains and layers of paint have distinctive features. Fibres have different colours, patterns or textures.

Student Book
pages 122–123

5.10 Modern analytical instruments

- Modern high-tech analytical equipment allows scientists to get **precise** and **reliable** (trustworthy) results from complex mixtures and tiny quantities of materials.
- Analytical techniques (to identify samples such as drugs) include gas-liquid chromatography, mass spectrometry and infrared spectroscopy.

> **1** *Why is the graph below labelled as a 'positive urine test'?*

Key points

- Computers speed up forensic searches and make high-tech analysis of samples more precise and reliable.
- An unknown substance can be identified by matching the distinctive features of its trace with those of known substances on a computer database. [H]

AQA *Examiner's tip*

In the exam you do not need to know details of the analytical instruments, but you will be expected to interpret traces to identify a match between an unknown and known substance.

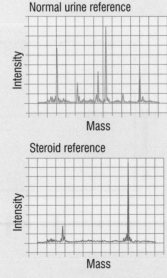

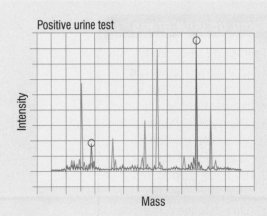

Mass spectrometry data from an athlete's sample

Key words: precise, reliable

Higher

The use of science in analysis and detection

Student Book pages 124–125

5.11 Blood and DNA

- Blood contains red blood cells, white blood cells, platelets and plasma.
- The four main blood groups are A, B, AB and O.

Key points

- Blood contains red blood cells, white blood cells, platelets and plasma.
- DNA is found in cell nuclei. You inherit DNA from your parents.

⏩ **1** *What does blood contain?*

⏩ **2** *A suspect has blood group AB. How could this eliminate him from a police enquiry?*

- Coiled inside the **nucleus** of the cells of our bodies are spirals of DNA.
- Except for identical twins, the DNA of every individual is **unique**.
- You can extract samples of DNA from blood, saliva, semen and body cells.

Key words: nucleus

Student Book pages 126–127

5.12 DNA profiling

In **electrophoresis**, negative **DNA** fragments in the alkaline gel are attracted to the positive electrode. Smaller fragments move faster, but friction causes the less mobile, larger fragments to stick to the gel.

Key points

- Electrophoresis is like electrically forced chromatography. It separates DNA fragments to produce a DNA profile. [H]
- While some oppose our personal information being stored on national databases, others argue that it is only criminals who have something to fear.

- The pattern of bands on the gel is transferred by computer, so matches can be made to other profiles held on the national database.

⏩ **1** *How do the DNA fragments move through the alkaline gel in electrophoresis?* [H]

⏩ **2** *Explain the ethical implications of having a National DNA Database.*

Bump up your grade

To improve your grade:
- Use correct scientific terms. Know what key words mean.
- Make correct deductions from data provided.
- Give enough detail (in bullet points) for 'describe' questions.
- Do not leave questions blank, especially for 'suggest' questions.
- Give reasons for the facts in 'explain' questions.

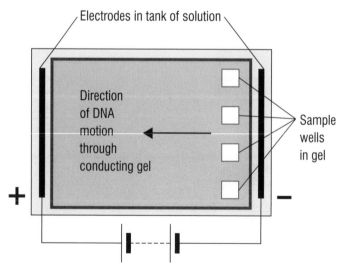

Electrophoresis

Key words: electrophoresis, DNA

5.13 Glass – the invisible evidence!

- Different types of glass each have a different **density**, so they change the direction of light by different degrees and have different **refractive index** values.

 Maths skills

We can find the refractive index by using the formula:
$$\text{refractive index} = \sin i \div \sin r$$
i is the angle of incidence and r is the angle of refraction.
e.g. If $i = 75°$ and $r = 40°$ then

$$\text{refractive index} = \frac{\sin i}{\sin r} = \frac{\sin 75°}{\sin 42°} = \frac{0.966}{0.669} = 1.44$$

Remember your scientific calculator for the exam.

1 *If the angle of incidence is 45° and the angle of refraction is 28°, what is the refractive index of a glass block?*

- To measure the **refractive index** of a tiny bit of glass, immerse the fragment in an oil, under a microscope. Slowly heating and cooling the oil changes its density. At a certain temperature the glass seems to disappear – when the **refractive index** of glass and oil match. A computer program can then convert the temperature into a refractive index value.

2 *What does a forensic scientist do to make it seem that a glass fragment disappears?*

Key words: refractive index

Student Book pages 128–129

Key points

- Light refracts (changes direction) towards the normal when it enters glass or plastic.
- We can find the refractive index by using the formula: refractive index = sin i ÷ sin r
- Oil immersion method: a glass fragment disappears when put in oil at a certain temperature with the same refractive index.

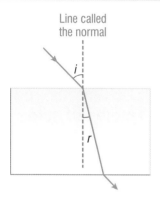
Line called the normal

Refraction of light by a glass block

5.14 Standard procedures for analysis

- Standard procedures for analysis include using limewater to test for carbon dioxide and acidified potassium dichromate to test for ethanol.
- Flame tests can detect metal ions and universal indicator can detect acidity.
- In titrations scientists analyse amounts of chemicals in solutions.
- Chromatogram retention factor calculations help identify different chemicals dissolved in solution.
- Refractive index calculations help identify plastics and glasses.
- To remove insoluble solid matter, filter the mixture through **filter paper** to collect a clear solution.
- To test for **solubility** evaporate 100 g of a saturated solution over a water bath and calculate the mass of the residue. The solubility is the number of grams of compound dissolved per 100 g of water.

Evaporating water from a solution

1 *If 7 g of a substance forms a saturated solution in 25 g of water, what is its solubility?*

Key words: solubility

Student Book pages 130–131

Key points

- Standard procedures are used to test for the presence, amount and composition of a chemical.
- Refractive index calculations help identify plastics and glasses.
- Solubility is the number of grams of compound dissolved per 100 g of water.

Bump up your grade

You can expect questions on standard procedures, so learn the steps in the standard procedures for your examination.

1 What is a precipitate?

2 Why do analytical scientists use flame tests?

3 What was the chemical that changed colour in the original design of breathalyser?

4 State three ways to distinguish one pollen grain from another.

5 When does a glass fragment seem to disappear in the 'oil immersion method'?

6 Why do ionic compounds have higher melting points than most covalent compounds?

7 Why do some chemicals travel shorter distances than others in paper chromatography? [H]

8 In the acid–base titration: $HCl + NaOH \longrightarrow NaCl + H_2O$, if 0.1 moles of acid are used, what mass of NaOH is used to neutralise it?
(Relative atomic masses: H = 1, O = 16, Na = 23)

9 **a** Write a balanced symbol equation for the reaction:

hydrochloric acid + calcium carbonate $\longrightarrow$ calcium chloride + water + carbon dioxide [H]

b Show that if 146 g of HCl are added to excess calcium carbonate, 88 g of carbon dioxide are generated. [H]
(Relative atomic masses: H = 1, C =12, O = 16, Cl = 35.5)

Chapter checklist ✓✓✓

Tick when you:

reviewed it after your lesson	✓	☐	☐
revised once – some questions right	✓	✓	☐
revised twice – all questions right	✓	✓	✓

Move on to another topic when you have all three ticks

✓✓✓

Introduction to analytical science	☐	☐	☐
Distinguishing different chemicals	☐	☐	☐
Testing for ions	☐	☐	☐
Breathalysers	☐	☐	☐

Flame tests	☐	☐	☐
Balanced equations	☐	☐	☐
Titrations	☐	☐	☐
Chromatography	☐	☐	☐
Microscopic evidence	☐	☐	☐
Modern analytical instruments	☐	☐	☐
Blood and DNA	☐	☐	☐
DNA profiling	☐	☐	☐
Glass – the invisible evidence!	☐	☐	☐
Standard procedures for analysis	☐	☐	☐

1 Sodium chloride (common salt) is often added to foods to improve their taste. Food scientists regularly test foods for their levels of salt content.

  a What type of chemical bonding exists in sodium chloride? *(1 mark)*

  b When sodium reacts with chlorine, the sodium forms a 1+ ion; the chlorine forms a 1– ion. What is the formula of sodium chloride? *(1 mark)*

  c Sodium chloride forms a giant lattice structure, as follows:

  Explain why sodium chloride has a very high melting point. *(2 marks)*

  d Describe a technique a food scientist could use to test for the presence of sodium ions. Include a safety precaution the scientist should take when performing this procedure. *(4 marks)*

2 A forensic scientist wishes to identify a piece of glass found at the scene of a crime.

  a Copy and complete the diagram below, labelling the angle of incidence and the angle of refraction:

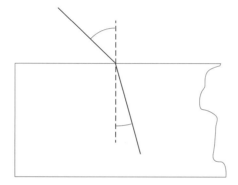

  (1 mark)

  b The following measurements were taken by the forensic scientist:

  Angle of incidence = 48.0°   Angle of refraction = 27.7°

  Using the table below, identify the type of glass found at the scene of the crime.

  Refractive index $= \dfrac{\sin i}{\sin r}$

Type of glass	Refractive index
Acrylic	1.49
Crown	1.52
Flint	1.60
Pyrex	1.47

  (3 marks)

  c Describe one way the scientist could ensure that her results are reliable. *(1 mark)*

  d Describe a standard procedure for measuring the refractive index of a glass fragment. *(4 marks)*

Answers

1 Working safely

1.1–1.3
1 Not cooperating with health and safety puts you and others at serious risk.
2 a warn of danger – yellow and black triangles
  b 'must not do' – red crossbar
  c 'must do' – blue circle
  d safety information – green background
  e hazard – orange and black
3 A hazard is a danger or anything that can cause harm. Its risk is the chance that someone might be harmed by that hazard.
4 a Water extinguisher
  b Carbon dioxide extinguisher. (You would not want powder getting inside the computers.)

1.4
1 So each time a test is done it is known to be reliable. So public safety is not compromised.
2 A hypothesis is an educated or scientifically reasoned guess answering the question that has been posed.

1.5
1 Standard procedures ensure everyone carrying out a task does it the same way. This limits human error and different approaches, so results should always be consistent no matter who carries out the procedure.

2 The use of science in maintaining health and fitness

2.1
1 Ventricles

2.2
1 Contracts and moves down
2 TV refers to the volume for normal breathing, while VC refers to the maximum volume.

2.3
1 They increase.
2 glucose + oxygen $\longrightarrow$ carbon dioxide + water (+ energy)
3 Oxygen debt occurs during vigorous exercise when your heart and lungs cannot send oxygen to your muscles quickly enough and anaerobic respiration takes over.

2.4
1 Insulin
2 Glycogen
3 Blood glucose levels fall. The pancreas releases glucagon. The liver converts glycogen to glucose.

2.5
1 So the body functions at its best
2 Sweat glands
3 You sweat and blood vessels supplying the capillaries expand (dilate) meaning more blood travels near to the skin's surface.

2.6
1 Tendons join muscle to bone. Ligaments join bone to bone.
2 Synovial fluid

2.7
1 Your biceps relax and triceps contract.
2 Moment = force × distance
  $2.4 = 80 \times B$
  $B = 2.4 \div 80 = 0.03\,m$ or $3\,cm$

2.8
1 Light so it doesn't limit acceleration, strong so it doesn't break and elastic to transfer energy and give a spring in the step

2.9
1 Glucose
2 You may forget the food you ate or have disposed of the food labels.

2.10
1 BER = $50\,kg \times 5.4\,kJ$ per hour
  = $50 \times 5.4 \times 24\,kJ$ per day
  = $6480\,kJ$
2 a BMI = $m \div h^2 = 80 \div 1.8^2 = 80 \div 3.24 = 24.7$; round up to $25\,kg/m^2$.
  b Being overweight increases your risk of heart disease and other illnesses.
  Reducing your carbohydrates and increasing your exercise will help.
3 An athlete with a muscular build and high BMI may not be fat or overweight.

2.11
1 a To replace water lost when sweating during exercise
  b To give an energy boost
  c To aid absorption of the drink into the blood stream and reduce urine output
2 Hypertonic
3 In order to build muscle, as sprinters are muscular powerful athletes, who use lots of energy every second

2.12
1 Choose from the bullet points e.g. Doctors help prevent, diagnose and treat medical conditions.
2 A sudden movement could give an inaccurate force reading.

Answers to end of chapter answers

1 To provide them with the energy to exercise
2 Your biceps contract, your triceps relax. (Skeletal muscles are antagonistic pairs.)
3 Pulmonary
4 Glucagon
5 Water, glucose, sodium/potassium salts
6 BER = $80 \times 5.4 \times 24 = 10368\,kJ$
7 Intercostal muscles and the diaphragm
8 a Tendons attach muscles to bone and make them move.
  b Ligaments hold bones together and keep joints stable.
9 When you are hot:
  – The energy taken from your body to evaporate the water in sweat causes cooling.
  – The diameter of your blood vessels supplying the capillaries in your skin increases, so more blood flows near the surface of the skin. Heat is transferred away from the skin by conduction, convection and radiation.
10 Oxygen debt occurs when your heart and lungs cannot supply enough oxygen to your muscles for aerobic respiration and anaerobic respiration takes over, causing a build up of lactic acid, which makes your muscles ache, and has to be turned back into glucose.

Answers to examination-style questions

1 a *1 mark* for two correctly matched, *2 marks* for three correctly matched, *3 marks* for all correctly matched.
  Carbohydrates – Main source of energy
  Proteins – Used for growth and repair
  Vitamins and minerals – maintain health
  Fats – Act as a store of energy, and are used for insulation
  b Glucose + oxygen $\longrightarrow$ (1 mark)
    carbon dioxide + water (1 mark)
  c Rises after eating (1 mark)
    Falls during exercise (1 mark)
  d Marks awarded for this answer will be determined by the quality of written communication (QWC) as well as the standard of the scientific response.
  There is a clear, detailed description of how the body responds to a change in blood glucose levels during exercise. Details should include references to hormones made. The answer shows almost faultless spelling, punctuation and grammar. It is coherent and in an organised, logical sequence. It contains a range of appropriate or relevant specialist terms used accurately including: insulin, glucagon, glycogen, and hormone. (5–6 marks)
  There is a relevant description of how the body responds to a decrease in blood glucose level. The answer has some structure and organisation. There are some errors in spelling, punctuation and grammar. The use of specialist terms has been attempted, but not always accurately. (3–4 marks)
  There is a brief description of how the body responds to an increase in blood glucose level. The answer is poorly organised with poor spelling and grammar. Almost no specialist terms are used and/or their use demonstrating a general lack of understanding of their meaning. (1–2 marks)
  No relevant comments (0 marks)
  Examples of the points made in the response:
  • How the body responds to both an increase and decrease in blood glucose level, and the effect on the body of these changes being made.
  • When blood glucose level rises pancreas secretes insulin.
  • Insulin causes liver to convert soluble glucose into glycogen.
  • The glycogen is stored, so blood glucose level falls to within normal limits.
  • When blood glucose level falls, pancreas secretes glucagon.
  • Glucagon causes liver to convert glycogen into glucose.
  • Glucose is then released back into bloodstream, so blood glucose level rises to within normal limits.
2 a (Heart) beats per minute/number of times heart beats per minute (1 mark)
  b Temperature (increases); breathing rate (increases) (2 marks)
  c Up to 2 marks for effect; up to 2 marks for explanation.
  Blood vessels supplying the capillaries widen; more blood flows closer to the surface of the skin allowing energy to be transferred (heat to be lost) by radiation.
  Sweating; as water in sweat evaporates energy (heat) is removed from the surface of the skin.

Hairs on the body lie flat; skin becomes less insulated.
(Accept the reverse answers.) *(4 marks)*

d Oxygenated blood returns from the lungs to the left atrium.
Left ventricle pumps oxygenated blood to the rest of the body.
Deoxygenated blood returns to the right atrium.
Valves prevent blood flowing backwards through the heart.
(4 marks)

3 a Any two from: damaged ligaments; pulled/torn muscles; torn cartilage; ruptured tendons; dislocated joints; fractured/broken bones; reference to a specific ligament/joint/bone *(2 marks)*

b Pulling
Contracting *(2 marks)*

c i Moment = 4.5 × 40 *(1 mark)*
= 180 N cm or 1.8 N m *(1 mark for value, 1 mark for unit)*

ii 240 (N cm) or 2.4 (N m) *(1 mark)*

iii Total moment required =
180 + 240 = 420 N cm or 4.2 N m

moment = force × distance;
force = $\dfrac{\text{moment}}{\text{distance}} = \dfrac{420}{5}$ or $\dfrac{4.2}{0.05}$
= 84 N *(3 marks)*
Award one mark for correct working in **c i** and **iii** if final answer is incorrect.

3 The use of science to develop materials for specific purposes

3.1

1 a British Standards Institute
b European Committee for Standardisation

2 Volume = 1 cm × 2 cm × 5 cm = 10 cm³
Density = mass ÷ volume = 45 g ÷ 10 cm³ = 4.5 g/cm³

3.2

1 a Radius = 1 mm
Area = $\pi \times r^2 = \pi \times 1^2 = 3.14 \text{ mm}^2$
b Stress = force ÷ area = 1000 N ÷ 3.14 mm²
= 318 N/mm²
So tensile breaking strength ≈ 320 N/mm²

3.3

1 E.g. oil platform
2 It is hard to break the bonds that hold the positive ions and electrons together.

3.4

1 Their high melting point and strength stops them melting or bending due to friction when braking. Their light weight helps the car accelerate faster (than with steel disc brakes).

3.5

1 a Their flexibility allows joints to bend freely.
b Their elasticity grips muscles to encourage blood flow.
2 a Toughness ensures they absorb energy (rather than shattering) when hit by stones or bullets or knives.
b Strength ensures the artificial ligament does not tear.

3.6

1 They have few cross-links, if any. Their relatively weak intermolecular bonds allow them to deform and flow.

3.7

1 a Tennis rackets, made with a composite of titanium and carbon/Kevlar fibres in resin, are lightweight (with a low density) to aid manoeuvrability and tough to absorb energy and dampen vibrations.
b Support stockings, made with Lycra, are elastic and tight-fitting to encourage blood flow to muscles.
c Aircraft frames, made with aluminium alloy, have a low density, so are lightweight and easier to accelerate on take-off, so save on fuel.

3.8

1 Tough
2 a Radius = 1.5 mm
$A = \pi r^2 = \pi \times 1.5^2 = 7.07 \text{ mm}^2$
b Stress = force ÷ area = 540 kN ÷ 7.07 mm²
= 76 kN/mm²

Answers to end of chapter questions

1 When it is stretched beyond its elastic limit.
2 Tensile implies stretching or pulling apart. Compressive implies squashing or pushing together.
3 a Stiff
b Tough
4 An artificial hip joint
5 Lower density, does not corrode.
6 High melting point, resists chemical attack.
7 Density = mass ÷ volume
= 27 g ÷ 10 cm³
= 2.7 g/cm³
8 a Density
b Electrical resistance
c Spring constant
d Stress
9 a The sulfur creates strong intermolecular bonds or cross-links that branch between the polymer chains.
b This makes it rigid once set and both tough and strong, making it useful for car tyres and the soles of shoes.

Answers to examination-style questions

1 a The object is suitable for the purpose for which it is intended. *(1 mark)*
b BSI, CEN *(2 marks)*
c i Radius = 6 cm
Volume = 4/3 × 3.14 × 6³
Volume = 904.8 (cm³) *(3 marks)*
ii Mass = 8.02 × 904.8 *(1 mark)*
– error carried forward
Mass = 7256 g = 7.26 kg *(1 mark)*
Ball does meet the Olympic standard *(1 mark)*
(No mark is available for the final marking point without justification.)

2 a A long-chain molecule containing many repeating units (monomers) *(1 mark)*
b POLYETHENE –
Strong, low density, flexible –
Carrier bags
PVC –
Durable, tough, electrical insulator –
Insulation for electrical cables
PERSPEX –
Transparent, hard, tough –
Unbreakable 'glass' for watches

PET –
Transparent, flexible, low density –
Food containers
All four correct *(3 marks)*
Three correct *(2 marks)*
Two or one correct *(1 mark)*
c Thermoplastic – becomes flexible/softens on heating. *(1 mark)*
Thermosetting – can be moulded on first heating; does not then deform on subsequent heating. *(1 mark)*

3 a Two or more materials mixed together *(1 mark)*
b Marks awarded for this answer will be determined by the quality of written communication (QWC) as well as the standard of the scientific response.
There is a clear, detailed description of how each product has the identified material property linked to its application. The answer shows almost faultless spelling, punctuation and grammar. It is coherent and in an organised, logical sequence. It contains a range of appropriate or relevant specialist terms used accurately, including: (tensile) (compressive) strength, toughness, brittleness, density, hardness, melting (boiling) point, corrosion resistance, thermal (electrical) conductivity. *(5–6 marks)*
There is a description of two relevant material properties for each product. There are some errors in spelling, punctuation and grammar. The answer has some structure and organisation. The use of specialist terms has been attempted, but not always accurately. *(3–4 marks)*
There is a brief description of a property of two relevant products. Spelling, punctuation and grammar are very weak. The answer is poorly organised with almost no specialist terms and/or their use, and demonstrating a general lack of understanding of their meaning. *(1–2 marks)*
No relevant comments *(0 marks)*
Examples of the points made in the response:
- Mountain bike frames; low density, high toughness; low density makes the bike lighter, high toughness means the frame is less likely to break if a crack occurs in the frame.
- Racing car disc brakes; high strength, very hard; high strength means won't break, hardness means won't wear easily, so will last a long time.
- Aluminium–copper alloys for electrical cables; low resistivity, low density; low resistivity means current conducted efficiently, low density means cables have relatively low mass per unit length.
- Kevlar composite for bullet-proof jackets; tough, low density; tough means that if a crack starts in the material it will not propagate, low density means that the jacket is lightweight for the wearer.
- Stainless steel for saucepans; good thermal conductivity, high corrosion resistance; good conductivity means heat reaches food quickly and efficiently, high corrosion resistance will not react with oxygen/water/foodstuffs.
- Magnesium alloy for car alloy wheels; low density, high strength; low density means mass of car minimised, so more fuel efficient/better acceleration, high strength means does not break under extreme loading (e.g. in an accident).

4 The use of science in food production

4.1
1 Defra, FSA
2 Any three from: analysing food quality, checking food safety, detecting the presence of additives, research into producing new food products and better storage and transportation, preparing special diets.

4.2
1 *Campylobacter, Salmonella, E. coli*
2 Warmth, moisture and a food source

4.3
1 Any three from: using detergents, disinfectants, sterile packaging and equipment, appropriate waste disposal and adequate pest control
2 Any three from: hair net, disposable gloves, apron, hat, white coat
3 Any three from: heating, cooking, freezing, drying, salting, pickling, refrigerating

4.4
1 To allow time for the microorganisms to grow.
2 They are too small to see.

4.5
1 A good supply of glucose, with no oxygen present, and at a temperature between 15 °C and 25 °C.
2 It evaporates when the bread is baked.

4.6
1 Bacteria and rennet
2 Curdles the milk and hinders the growth of harmful bacteria

4.7
1 a Intensive: Higher yields/cheaper products
  b Organic: Animals have a better quality of life/ people believe food is healthier and tastier
2 Using natural predators to destroy pests

4.8
1 Build chemical plants near to raw materials, recycle any chemical by-products, ensure a good rate of chemical reaction and maximise the yield from a reaction.

4.9
1 Advantages – safe from predators/cheaper meat and products. Disadvantages – lower quality of life
2 Advantages – better quality of life. Disadvantages – more chance of being eaten, more expensive product

4.10
1 Water, wind
2 Growing one crop on an area of land

4.11
1 It would wilt.
2 Leaves would be pale and yellow.

4.12
1 Concentration, temperature, surface area, use of catalysts

4.13
1 Theoretical yield is the maximum amount of product which could be made. Actual yield is the amount of product that is actually made.

4.14
1 High temperature, high pressure and use of a catalyst

4.15
1 Advantages – any two from: higher yields, products available for longer, more uniform crop, extend an organism's tolerance range.

Disadvantages – any two from: reduces variation, reduces gene pool, some pedigree animals have poorer health, plants produced without seeds cannot reproduce.
2 Faster (occurs within one generation) and more accurate

4.16
1 Quantitative test

Answers to end of chapter questions

1 a Fermentation
  b Carbon dioxide and alcohol
2 a Bacteria
  b One from *Campylobacter, Salmonella, E. coli*
  c Sickness, diarrhoea and vomiting
3 a Always growing one crop on an area of land
  b Any two from: the same nutrients constantly being removed from the soil, leading to greater use of fertilisers; crop pests being concentrated in one area, leading to greater use of pesticides; reduction of biodiversity, because the habitat for some animals is removed as only one plant is grown.
4 Select hen which lays the most eggs and mate with best male.
Choose hen which lays the most eggs from the offspring, and mate with best male.
Repeat over many generations.
5 Working aseptically, use a sterilised wire loop to make streaks on an agar plate. Flame the loop, and allow to cool. Spread the bacteria about by streaking the plate in another direction. Repeat twice more. The species of bacteria can be identified by the appearance of its colonies.
6 a A reaction that can proceed in the forward or backward direction
  b $N_2 (g) + 3H_2 (g) \rightleftharpoons 2NH_3 (g)$
  c High temperature – to increase rate of reaction (but reduces yield); high pressure – to increase rate of reaction and yield; catalyst – to increase rate of reaction (has no effect on yield)
7 a Altering an organism's genes to produce desired characteristics
  b Frost-resistant tomato – can survive cold conditions
GM cotton – pest and disease resistant
Genetically engineered bacteria – produce insulin which can be used by diabetics.

Answers to examination-style questions

1 a One from:
    to detect the presence of microorganisms; determine the number of microorganisms in a food product; in the manufacture of [cheese/yoghurt/bread/beer/wine]. *(1 mark)*
  b Bacteria, cheese, yeast, bread *(4 marks)*
  c Fermentation *(1 mark)*
  d Glucose; ethanol (alcohol) and carbon dioxide *(2 marks)*
  e Sterilise milk
    bacteria added – fermentation occurs
    lactic acid is produced – lactic acid curdles milk
    lactic acid prevents harmful bacteria reproducing. *(4 marks)*
2 a i Intensive farming: the growth of crops and animals is maximised, resulting in high yields. *(1 mark)*
    ii Organic farming: natural methods of rearing animals and growing crops are used. Yields tend to be lower. *(1 mark)*

b 1 mark awarded for each correct row:
(4 marks)

Factors that can be controlled	Intensively reared animals	Organically reared animals
Temperature	Animals are kept indoors, in a warm environment.	Animals normally outdoors. Animals use energy to heat their bodies.
Space	Movement restricted to ensure energy used for growth.	Animals are allowed to roam as freely as possible.
Use of drugs	Antibiotics are given to prevent the spread of disease.	Antibiotics only given if an animal is ill.
Enclosure Safety	Animals are kept inside, safe from predators.	Live outdoors, so higher risk from predators.

c i Any one from: more space needed; more expensive feed required; slower growth rate; more labour intensive *(1 mark)*
  ii Because organic farming is a more natural process, animals able to exhibit natural behaviours
    So animals have a better quality of life/ animals experience less suffering
    Or
    meat contains fewer/no artificial chemicals.
    So the meat tastes better/healthier.
    (2 marks)

d Marks awarded for this answer will be determined by the quality of written communication (QWC) as well as the standard of the scientific response.
There is a clear, balanced and detailed description of how crops are produced organically. The answer shows almost faultless spelling, punctuation and grammar. It is coherent and in an organised, logical sequence. It contains a range of appropriate or relevant specialist terms used accurately, including: fertiliser, pesticide, herbicide, fungicide, biological control, nutrient, competition. *(5–6 marks)*
There is a description of how crops are produced organically. There are some errors in spelling, punctuation and grammar. The answer has some structure and organisation. The use of specialist terms has been attempted, but not always accurately.
(3–4 marks)
There is a brief description of how crops are produced organically which has little clarity and detail. The spelling, punctuation and grammar are very weak. The answer is poorly organised with almost no specialist terms and/or their use demonstrating a general lack of understanding of their meaning.
(1–2 marks)
No relevant comments *(0 marks)*
Examples of the points made in the response:
• Nutrient-poor soil; use of natural fertilisers/ manure; returns nutrients to soil
• Crop pests; use of biological control; predator removes pest from crops

- Weeds compete for nutrients/competition; weeding (by hand or machine); removes weeds from crops
- Fungal infection; removal of affected crop; minimises risk of infection spreading

5 The use of science in analysis and detection

5.1
1 DNA, hair, glass fragments, etc.
2 Defra helps the government to make policies and write laws covering environmental issues.

5.2
1 a CuS
  b $CuSO_4$
  c Na_2CO_3
  d $Pb(NO_3)_2$
2 a Ionic
  b Low

5.3
1 a Calcium hydroxide
  b Calcium
  c $Ca(OH)_2$
2 a Insoluble
  b Solid

5.4
1 a Accuracy to avoid a false reading.
  b Response time to avoid a delay before the reading is obtained.

5.5
1 Wear eye protection.
2 A dominant colour may prevent you identifying another metal ion present in the mixture.

5.6
1 $1 + 14 + (16 \times 3) = 63$, so 1 mole of $HNO_3 = 63g$
2 a Carbon dioxide
  b nitric acid + calcium carbonate $\longrightarrow$ calcium nitrate + water + carbon dioxide
  c $2HNO_3 + CaCO_3 \longrightarrow Ca(NO_3)_2 + H_2O + CO_2$

5.7
1 a $(V \div 1000) \times moles/dm^3 = (25 \div 1000) \times 0.2 = 0.005$
  b 0.005
  c NaOH: $23 + 16 + 1 = 40$
  d moles used × relative formula mass of alkai $= 0.005 \times 40 = 0.2g$

5.8
1 R_f = distance travelled by substance ÷ distance travelled by solvent = $3 \div 10 = 0.3$
2 When the mixture doesn't dissolve in water, thin layer chromatography is more appropriate.

5.9
1 A comparison microscope
2 A scanning electron microscope
3 Paint flecks

5.10
1 Two peaks occur on the trace of the athlete's sample which correspond to the trace of the steroid showing steroid is present – so a positive result.

5.11
1 Blood contains red blood cells, white blood cells, platelets and plasma.
2 If the evidence from the crime scene is only A, B or O, then the suspect was not present.

5.12
1 The DNA fragments in an alkaline solution are negatively charged, so are attracted towards the positive electrode.
2 While some oppose our personal information being stored on national databases as an erosion of their civil liberties, others argue that it is only criminals who have something to fear, and so the public should appreciate having such a database of information to identify law-breakers.

5.13
1 Refractive index = $\sin i \div \sin r = \sin 45° \div \sin 28° = 1.5$
2 Immerses it in oil, then heats and cools the oil until the glass fragment and the oil have the same refractive index.

5.14
1 If 7g dissolves in 25g of water then $7 \div 25g = 0.28g$ dissolve in 1g and $0.28 \times 100g = 28g$ dissolve in 100g so the solubility = 28g per 100g of water

Answers to end of chapter questions
1 An insoluble solid (in a liquid)
2 To identify metal ions
3 Acidified potassium dichromate
4 Three of: size, shape, colour or surface texture
5 When the refractive index of the glass fragment matches that of the oil (at a particular temperature).
6 You need more energy to break an ionic bond than separate molecules that are covalently bonded.

7 They cling more strongly to the paper (stationary phase) and are not so soluble in the water (which is the solvent or mobile phase).
8 To neutralise the acid there must also be 0.1 moles of NaOH.
40g of NaOH is 1 mole
so the mass of NaOH = $0.1 \times 40g = 4g$
9 a $2HCl + CaCO_3 \longrightarrow CaCl_2 + H_2O + CO_2$
  b Relative formula masses: HCl = $1 + 35.5 = 36.5$; $CO_2 = 12 + (16 \times 2) = 44$
  but 2HCl ($2 \times 36.5g = 73g$) generates 44g of CO_2
  therefore $2 \times 73g = 146g$ generates $2 \times 44g = 88g$

Answers to examination-style questions

1 a Ionic *(1 mark)*
  b NaCl *(1 mark)*
  c Many strong (ionic) bonds exist between (oppositely charged) ions
  So – A lot of energy (a high temperature) is required to separate the ions *(2 marks)*
  d Flame test
  Clean the wire in concentrated HCl.
  Sample is heated strongly in a (Bunsen) flame.
  Yellow flame indicates presence of sodium ions.
  Safety precaution: One from: wearing goggles/gloves/protective clothing/ appropriate warning signs displayed/care with flame or hot equipment *(4 marks)*
2 a Angle of incidence labelled between incident ray and normal line. Angle of refraction labelled between refracted ray and normal line. *(1 mark)*
  b $n = \sin 48.0°/\sin 27.7°$
  $n = 1.60$
  Type of glass is flint glass *(3 marks)*
  c Repeat measurements many times/using different apparatus/using different techniques *(1 mark)*
  d Immerse the glass fragment in a beaker of oil. Heat and/or cool the oil gently.
  When the glass fragment becomes invisible, note the oil temperature.
  Use a table of data/calibration graph/look up on database to convert the oil temperature to a refractive index. *(4 marks)*